Applied Surface Science

Edited by Gurrappa Injeti

Published in London, United Kingdom

IntechOpen

Supporting open minds since 2005

Applied Surface Science
http://dx.doi.org/10.5772/intechopen.78198
Edited by Gurrappa Injeti

Contributors
Rolando Roque Malherbe, Carlos De Las Pozas Del Rio, Umesh Vates, Maria Marinescu, Mariana Cabrera, David Neri, Fernando Soria, Luiz Carvalho Jr, Gulmira Yar-Mukhamedova, Marina Ved', Nikolay Sakhnenko, Alejandro Avila, Critian Carrera-Figueiras, Juan Antonio Juárez-Moreno, Yamile Pérez-Padilla, Erbin Guillermo Uc-Cayetano, Manuel Alejandro Estrella-Gutiérrez

First published in London, United Kingdom, 2019 by IntechOpen
IntechOpen is the global imprint of INTECHOPEN LIMITED, registered in England and Wales, registration number: 11086078, The Shard, 25th floor, 32 London Bridge Street
London, SE19SG – United Kingdom
Printed in Croatia

British Library Cataloguing-in-Publication Data
A catalogue record for this book is available from the British Library

Additional hard and PDF copies can be obtained from orders@intechopen.com

Applied Surface Science
Edited by Gurrappa Injeti
p. cm.
Print ISBN 978-1-78984-097-1
Online ISBN 978-1-78984-098-8
eBook (PDF) ISBN 978-1-78984-745-1

For EU product safety concerns:
IN TECH d.o.o., Prolaz Marije Krucifikse Kozulić 3, 51000 Rijeka, Croatia, info@intechopen.com or visit our website at intechopen.com.

Meet the editor

Prof. Dr. Gurrappa is a senior scientist in the Defence Metallurgical Research Laboratory, Hyderabad, India. He has worked in the area of materials science and nanostructured coatings for more than three decades. Apart from defense, he has been helping different industries to solve corrosion problems by stressing the need to prevent corrosion by adopting different advanced protective techniques. Prof. Gurrappa is recognized globally and has visited Germany, Japan, France, the USA, the UK, Belgium, the Netherlands, Poland, Singapore, Israel, and Malaysia. He has authored more than 200 publications both internationally and nationally. He delivered a number of invited lectures/talks and keynote addresses in various international/national symposia and has chaired technical sessions. Prof. Gurrappa is a recipient of prestigious fellowships across the globe in recognition of his meritorious contributions. He is also a recipient of the Corrosion Awareness Award and Meritorious Contribution Award by NACE International, India. He has published 11 books/book chapters and edited two books on gas turbines. He is an elected Fellow of the Royal Society of Chemistry, London; Andhra Pradesh Akademi of Sciences; and Founder Fellow of the Telangana Academy of Sciences, India.

Contents

Preface

Modern systems necessitate the use of advanced materials with appropriate surface properties essential to enhance their performance significantly. A surface that is exposed directly to different environments depending on the application is paramount because it needs to exhibit excellent resistance against wear, erosion, corrosion, temperature, pressure, etc. to improve durability and achieve ever-greater efficiency. Therefore, there is a need to obtain an appropriate surface finish depending on the application, which necessitates the usage of multiple surface engineering techniques that are complementary to each other in association with suitable surface preparation processes. To ensure the expected behavior and efficiency, modern experimental methods to assess surface characteristics as well as interfaces are of great importance. This book presents recent advances in the field of surface science for a better understanding of surfaces and interfaces and their applications. It is a useful book describing a variety of materials for different applications and gives an in-depth understanding of mechanisms involved in achieving the desired properties that are extremely useful to materials developers, design engineers, scientists, system maintenance engineers, academia, students, researchers, and industrialists. The book comprises a collection of articles written by senior people specialized in surface science and technologies.

Chapter 1 covers organic semiconductors as active materials in thin film electronic devices, such as alkynes, heterocyclics, dyes, ferrocenes, spiranes, or porphyrins, with special geometries and certain electronic molecular parameters that possess nonlinear optical properties and offer several major advantages over their inorganic counterparts. There are a number of simple and versatile techniques that can be employed for the deposition of these important classes of materials. The matrix-assisted pulsed laser evaporation (MAPLE) technique provides advantages with regard to making organic films of different morphologies on various types of substrates. New insights into crystallization growth mechanisms in MAPLE-deposited conjugated polymer films, which realize the connection between the structure and the carrier transport properties, are discussed herein. Second harmonic generation capabilities of thin films are also investigated.

Chapter 2 demonstrates different proposals of magnetic bioderivatives and their applicability in biotechnology. The historical context of immobilized enzymes, as well as immobilization methods highlighting the main advantages and disadvantages of each, is mentioned. In addition, iron oxides and composite materials are presented as support for biomolecule immobilization. Composites are effortlessly prepared, including many materials capable of providing advantages to the magnetic derivatives. Enzymes covalently linked to these magnetic particles combine their catalytic properties with reaction specificity, reusability, and possible reactor construction. In addition, proteins can also be purified by these magnetic composites containing specific ligands allowing reactors and reuses too. A number of characterization techniques used to study the magnetic material and derivative immobilized are described as well. Altogether, an engaging presentation of the interesting features of magnetic bioderivatives is highlighted, as well as their uses in the biotechnology field and other applications.

Chapter 3 discusses surface science engineering through the sol-gel process and its significance. Sol-gel synthesis is used to obtain coatings that can modify the surfaces of metals to avoid corrosion or to enhance the biocompatibility and bioactivity of metals and their alloys that are of biomedical interest. Anticorrosion coatings composed of smart coatings and self-healing coatings are described. TiO_2, hydroxyapatite, bioglass, and hybrid coatings synthesized by sol-gel technology are briefly introduced with regard to their role in modifying the surface of metals for biomedical purposes. Finally, although there are other approaches to modify the surface of metals for either anticorrosion or biomedical purposes, sol-gel methods have several advantages in controlling surface chemistry composition and functionality.

Chapter 4 describes the synthesis, characterization, and adsorption properties of nanoporous materials: nitroprussides, Prussian blue analogs, akaganeites, MeAPOs, metal organic frameworks, and highly specific surface amorphous silica, which allow the storage of about 11 wt% of hydrogen in the form of ammonia. In this sense, the use of solid-state reaction method and sol-gel methodologies, together with aluminosilicate, high silica, and non-aluminosilicate zeolite synthesis methods, is described. Moreover, the preparation of active carbons along with the synthesis of Prussian blue analogs and nitroprussides is explained. In addition, characterization of the materials of interest applying X-ray diffraction, thermogravimetric analysis, DRIFTS, and room-temperature Mossbauer spectrometry is discussed. Besides, were defined the concepts which define physical adsorption, provided examples of adsorption data, which were tested with the help of the Dubinin, osmotic adsorption, and Langmuir-type isotherms. Additionally, the methodology for the measurement of adsorption data with the help of the volumetric method is described. Moreover, a description of the thermodynamics of adsorption, as well as the methodology for the calculation of calorimetric data with the help of heat flow calorimeters together with the measurement of differential heats of adsorption data, is analyzed. Finally, the different interaction forces that make adsorption possible are discussed.

Chapter 5 presents ternary alloys of cobalt with molybdenum and tungsten deposited from biligand citrate-pyrophosphate electrolytes by the pulsed mode, which exhibits different compositions and surface morphologies depending on current density and on/off time. It is shown that the current density of 910 A dm^2 and on/off time of 5/20 ms favor the maximum refractory metals content in the deposits, but increasing current density diminishes the efficiency of electrolysis. When depositing Co-Mo-W from the electrolyte, the ratio Mo:W is 1:2 and tungsten content in the alloy is found to be much lower in comparison with molybdenum: W 27 at% vs. Mo 16 22 at%; the total of Mo + W content is 1527 at%. Therefore, changing the concentrations of Mo and W salts in the electrolyte for Co-Mo-W alloy deposition does not influence Mo and W content in the deposit in the same proportions. It is established that a sufficiently uniform surface of binary Co-Mo and Co-W deposits with smaller spheroid sizes and without microcracks may be attributed to lower refractory metals content. The structure of binary and ternary alloys is found to be amorphous crystalline, and intermetallic phases Co_7W_6 and Co_7Mo_3 are identified in deposits. The coherent-scattering region size of the amorphous part is detected at 2–8 nm. The amorphous structure of ternary alloys and significant content of alloying elements (Mo and W) predetermine improved high corrosion resistance. Corrosion resistance of binary and ternary deposits increases with the total content of refractory metals, which, associated with molybdenum and tungsten, enhance corrosion resistance to pitting as well as decrease roughness and smooth out the relief of ternary coatings. Ternary galvanic alloys of cobalt with molybdenum and zirconium of different compositions and morphologies are obtained from polyligand citrate-pyrophosphate electrolytes in a pulsed mode by varying the current density and duration of pulse and pause. It is shown that coatings with microglobular morphology with a low level of stress and cracks are formed at a current density of 46 A dm^2 and polarization on/off time of 2/10 ms. High corrosion resistance of ternary coatings based on cobalt is caused by an increased tendency to passivity and high resistance to pitting corrosion in the presence of molybdenum and zirconium, as well as the acid nature of their oxides.

Chapter 6 describes the attempts made to investigate the experimental process and surface roughness (SR) optimization of cold working (high carbon and high omium) hard die steel (D2) during the wire electrical discharge machining (WEDM) processes. The WEDM process relies heavily on the operators' technologies and experience because of their numerous and diverse range. Because using complicated cuts can make it difficult to machine electrically conductive components, the WEDM process was developed to generate precise cutting on complicated and difficult to machine materials. Tan-sigmoid and purlin transfer function with a bias-based four-layered back propagation artificial neural network approach have been used to investigate the effect of six independent parameters, namely gap voltage (V_g), flush rate (F_r), pulse on time (T_{on}), pulse off time (T_{off}), wire feed (W_f), and wire tension (W_t) over the Center Line Average value of surface roughness (Ra) along with corresponding material removal rate (MRR). Model adequacy is most satisfactory because correlation coefficient (R^2) is found to be 99.1% and the adjusted ($R_{adj.2}$) statistics is 98.5. It was found that the spark time ON/OFF, wire feed rate, wire tension, gap voltage, flush rate and a few of their interactions have a significant effect on SR.

I strongly believe that this book will be useful to materials scientists, system design engineers, manufacturing experts and executives, mechanical engineers, chemical engineers, aeronautical engineers, academic researchers, undergraduate and postgraduate students, and industrialists.

Dr. Gurrappa Injeti
Professor,
Defence Metallurgical Research Laboratory,
Hyderabad, India

Chapter 1

Synthesis and Nonlinear Optical Studies on Organic Compounds in Laser-Deposited Films

Maria Marinescu

Abstract

Organic semiconductors as active materials in thin-film electronic devices such as alkynes, heterocycles, dyes, ferrocenes, spiranes, or porphyrins, with special geometries and certain electronic molecular parameters, which possess nonlinear optical (NLO) properties and offer several major advantages over their inorganic counterparts, are presented in this chapter. There are a number of simple and versatile techniques that can be employed for the deposition of these important classes of materials. The matrix-assisted pulsed laser evaporation (MAPLE) technique provides advantages with regard to making organic films of different morphologies on different types of substrates. New insights into the crystallization growth mechanisms in MAPLE-deposited conjugated polymer films, which realize the connection between the structure and the carrier transport properties, are discussed herein. Second harmonic generation (SHG) capabilities of the thin films were also investigated.

Keywords: organic synthesis, laser deposition, nonlinear optical properties, thin films

1. Introduction

During the last decades, the nonlinear optical (NLO) materials have gained significant role because of their various applications in medicine, molecular switches, luminescent materials, laser technology, spectroscopic and electrochemical sensors, data storage, microfabrication and imaging, modulation of optical signals, and telecommunication [1–3]. Organic materials are distinguished by the fact that they exhibit strong nonlinear optical (NLO) properties [4–8]. In the last years, researchers have based on the synthesis of the target organic molecules with particular geometries and certain electronic molecular parameters, in order to have the desired nonlinear optic properties [9–31].

The changes of optical properties (absorption coefficient, index of refraction), through the increasing intensity of the input light, led to the discovery of the nonlinear optical phenomenon, second harmonic generation (SHG), detectable only after the improvement of the laser in 1962 [32]. Thus, nonlinear optics developed as a tremendous field of research, especially after the profound understanding of nonlinear optic phenomena (NLO) and the structure-property relations of chromophores, after the development of different tools to accurately measure and calculate hyperpolarizabilities [33].

Recent literature highlights the increased interest in organic materials in recent decades, as an alternative to their inorganic counterparts, and having several advantages, such as their low cost, low toxicity, ease of solution processability, flexibility for device fabrications [34], and modulation of their optical, electronic, and chemical properties by adapting their molecular structure. Field effect transistors, photovoltaic devices, organic light-emitting diodes (OLEDs), and white light sources for indoor and outdoor lighting are some of the applications of organic materials [33].

The deposition of organic materials in thin films, required for the design of new, successful devices, implied the precise monitoring of their chemical, structural, and morphological properties [35]. The deposition of organic substances in thin films has to meet the requirements of the market: (1) good uniformity of simple or multilayer structures of organic, polymeric, or composite materials—in the electronics industry; (2) thickness control, film uniformity of coating, and good interfacing properties—in OLED polymer applications; (3) conformal coatings required to modify the interior surfaces of porous materials (membranes, foams, textiles) or irregular geometries of surfaces—for optoelectronic and medical devices [36].

Several classes of organic compounds, including conjugated molecules, fullerenes, polymers, perylenes, dyes, and thiophenes, have been studied as materials and investigated for their NLO responses [5]. Conjugated organic polymers with large nonlinear responses correlated with rapid response time have been observed as NLO materials with great expectations [37]. Although organic compounds have been considered as frail, the experiments showed, with the optical damage, threshold for polymeric materials can be greater than 10 GW/cm^2 [37].

Two deposition techniques, physical and chemical, are used in order to obtain *organic thin films* with good quality. For each type, there are several techniques applied. Physical deposition techniques for thin organic films include physical vapor deposition (PVD) [38–42], organic vapor phase deposition (OVPD) [43–45], organic molecular beam deposition (OMBD) [46–51], solvent vapor annealing (SVA) [52–56], self-assembled monolayers (SAMs) [57, 58], inkjet printing [59, 60], pulsed laser deposition (PDL) [61–64], and laser evaporation [65–67] techniques. The chemical methods include solution techniques and gas-phase deposition methods. Techniques that use solutions include Langmuir-Blodgett (LB) [68, 69], spin coating [70, 71], dip coating [72, 73], sol-gel [74, 75], and spray pyrolysis [76]. Chemical vapor deposition (CVD) [77–82] uses the gas phase of organic compounds.

Many articles report the synthesis of the novel organic molecules or polymers with highly active chromophores and superior optical activity, as response to the demand of substances with NLO properties for various applications [83–86].

This chapter refers to synthesis of organic compounds with nonlinear optical properties in one of the techniques mentioned above, laser-deposited films.

2. Nonlinear optical (NLO) response in organic molecules

The optical response is due to a transition of the dipole moment from the ground state to the excited state due to the transition of an electron between frontier orbitals, from the highest occupied molecular orbital (HOMO) to the lowest unoccupied molecular orbital (LUMO). The chemical activity of the molecule and the availability of the internal charge transfer are due to the balance between the redox ability of HOMO (as reducing agent) and LUMO (as oxidizing agent), which reveals the internal charge transfer responsible for the non linear optical properties. Nonlinear materials are defined as optical media in which the refractive index depends on light intensity [87]. So, the HOMO-LUMO gap energy is involved in molecular electrical transport properties.

The designing and obtaining (synthesis) of the new molecules with high first hyperpolarizability β (theoretical and experimental) is central in discovery of the second-order and higher-order nonlinear optical materials and is quantified by the induced dipole moment under an intense light field E in Eq. (1):

$$\mu_i = \alpha_{ij}E_j + \beta_{ijk}E_jE_k + \gamma_{ijkl}E_jE_kE_l + \ldots \tag{1}$$

with μ_i the ith component of the induced dipole moment, E_j the corresponding component of the applied electromagnetic field, and α_{ij}, β_{ijk}, γ_{ijkl}, the components of the linear polarizability, the first hyperpolarizability, and the second hyperpolarizability. In case of an ensemble of molecules, the macroscopic polarization is defined by Eq. (2):

$$P_i = \chi_{ij}{}^{(1)}\, E_j + \chi_{ijk}{}^{(2)}\, E_jE_k + \chi_{ijkl}{}^{(3)}\, E_jE_kE_l + \ldots. \tag{2}$$

with $\chi_s{}^{(1),\,(2),\,(3)}$ the macroscopic susceptibilities of the first (1), second (2), and third (3) order, which can be directly related to the density of the organic chromophore [88]. Recent advances in chromophore design report some features for classic dipolar organic structures with good nonlinear optic properties [89]: (1) presence of a π-conjugated systems with π electron delocalization, (2) a "push-pull" system, which is a couple donor-acceptor or connected to a system that contributes to the delocalization of the π electrons; (3) presence of a strong electron donor groups (e.g., $-NR_2$, $-NHR$, $-OR$, $-OH$), and strong electron withdrawing groups (e.g., $-CF_3$, SO_2CF_3, $-SO_3H$, $-NO_2$, $-CN$), positioned at opposite ends of a conjugated molecule in case of dipolar molecules; (4) great values of dipole moment and polarizability; (5) small HOMO-LUMO energy gap; (6) planarity of the molecule for neutral, polar, and zwitterionic resonance structures. Dipole organic molecules have an intrinsic matter: the dipoles prefer to align antiparallel with each other in the solid-state film to nullify the bulk effect. Octupolar molecules, alternative NLO materials, present more advantages compared with dipole molecules [90]: (1) the second harmonic response (SHG) does not depend on the polarization of the incident light because they are more isotropic than the dipolar molecules; (2) β values of the octupoles can be increased by increasing of intramolecular charge transfer; (3) octupoles form noncentrosymmetric crystals; and (4) they are less likely to undergo relaxation due to the lack of ground-state dipole moment.

3. Synthesis of the compounds with NLO properties

In the last decades, literature reveals some classes of organic compounds suitable for organic electronic devices, such as organic photovoltaics (OPVs) and organic thin-film transistors (OTFTs), which possess certain characteristics, such as high molecular hyperpolarizability coefficients (β), special geometry, and in most cases, small HOMO-LUMO energy gaps [25–27]. Among these classes of organic compounds, there are highlighting fullerenes, perylenes, thiophene compounds, polymers, and dyes.

Furthermore, the polymers represent one of the most used classes of substances in pulsed laser deposition (PLD), but also in the other methods for preparing thin films. Organic compounds with nonlinear optical properties and organic compounds reported in laser deposition (PLD) will be presented in the following two sections.

3.1 Synthesis of the nonlinear optical fullerenes

Canulescu group [91] studied thin films of fullerenes (C_{60}) **1** deposited onto silicon using matrix-assisted pulsed laser evaporation (MAPLE). MALDI analysis showed that a dominant transfer of intact C_{60} molecules onto a silicon wafer is realized when the laser fluence is carefully selected, for example, below a threshold of ~1.5≈/J cm^2. Labrunie et al. synthesized triphenylamine-based push-pull σ-C_{60} dyad, as photoactive molecular material for single-component organic solar cells, using a copper(I)-catalyzed 1,3-dipolar Huisgen cycloaddition under strict anaerobic conditions, leading to the selective formation of a 1,2,3-triazole ring and affording fullerene **2** in 80% yield (see **Figure 1**).

Thin films of **2**, prepared by spin coating of a $CHCl_3$ solution, highlight the ambipolar semiconducting behavior and also very good electron-transporting properties for fullerene **2** [92]. Kim et al. synthesized new fullerene **3** soluble in ethanol/water solvent mixtures and implemented these materials to fabricate polymer solar cells (PSCs) using environmentally friendly solvents [93]. The results of this paper provide important guidelines for the design of aqueous-electroactive materials having high carrier mobilities suitable to achieve very efficient eco-PSCs.

Kamanina highlights that there are two reasons for the importance of fullerenes: their unique energy levels and high value of electron affinity energy (0.65–0.7 eV). This value is larger than the one for most dyes and organic molecules with intramolecular acceptor fragment and can stimulate the efficient intermolecular charge transfer complex formation in the fullerene-doped organic conjugated materials [94].

Although fullerene acceptors were the predominant choice in the acceptor materials for two decades, the limited tunability of electronic properties and weak absorption of fullerene derivatives in visible range prevent further development of organic solar [95]. Therefore, other classes of organic molecules have been researched to obtain the desired properties of the electronic materials.

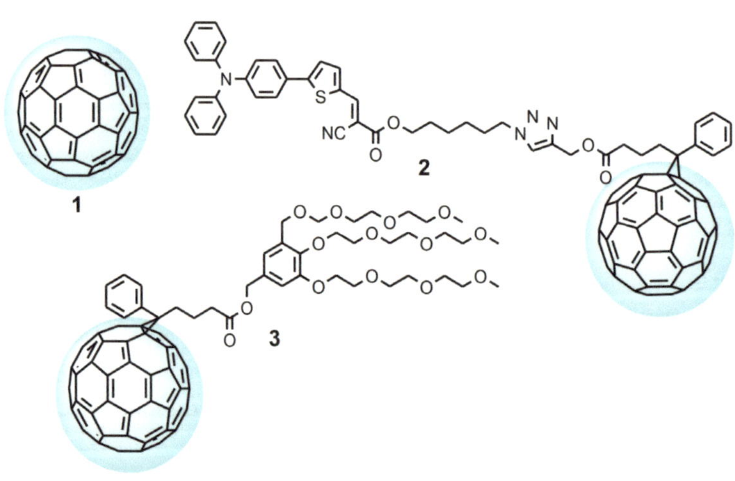

Figure 1.
Fullerenes with NLO properties [91–93].

3.2 Synthesis of the nonlinear optical perylenes

Perylene compound **4** was synthesized by a catalyzed heteroarylation reaction by McAfee et al. [83]. The geometry of perylene **4** exhibits planes of diketopyrrolopyrrole and perylene diimide at one dihedral angle too high for a good π-orbital overlapping, which determines HOMO-LUMO orbitals on specific atoms, supposition confirmed by TDDFT calculations at B3LYP/6-31G(d,p) level of theory. Optical, electronic, and self-assembly properties of the thin films of perylene **4** fabricated by solvent vapor annealing (SVA) recommended this compound as suitable for organic solar cells (OSCs) (see **Figure 2**).

Perylene diimides (PDIs) are a new class of nonfullerene electron acceptors for organic solar cells with many attracting features, like low cost; significant thermal, chemical, and light stability; good electron-accepting ability; and excellent electron mobility [96]. Carlotti et al. investigated PDI dimers as nonfullerene electron acceptors [96] for organic solar cells. Two isomers **5** and **6** have planar and twisted geometries, which determined very diverse spectral and photophysical properties (see **Figure 3**). Theoretical calculations and also the experimental time-resolved investigation confirm isomers **5** and **6** show charge transfer following light excitation.

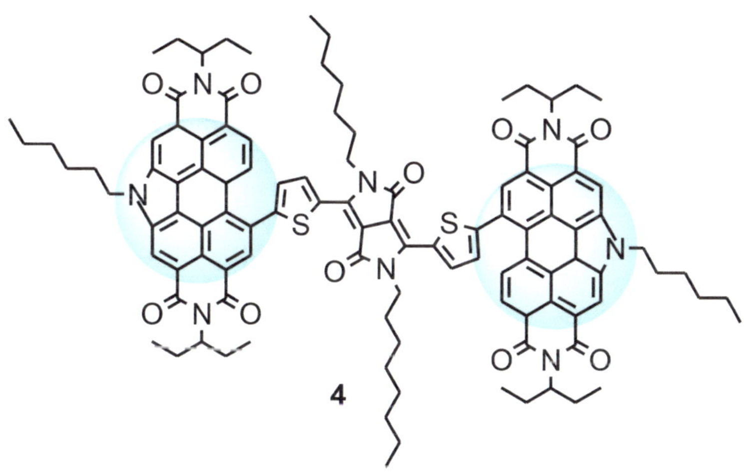

4

Figure 2.
*Perylene **4** with NLO properties [83].*

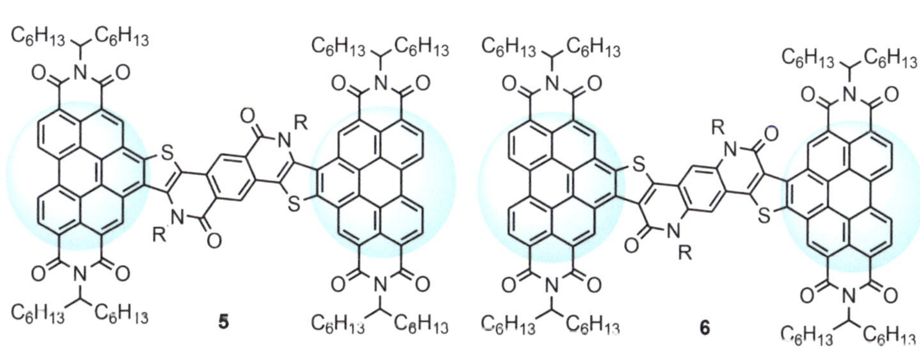

5 **6**

Figure 3.
*Perylene isomers **5** and **6** [96].*

3.3 Synthesis of the nonlinear optical thiophene compounds

Small-molecule semiconductors with an A-D-A core structure (D is an electron-rich unit and A is an electron-deficient unit) function as an electron donor or electron acceptor in organic photovoltaic cell devices [97].

Thiophenes are one of the most studied heterocyclic compounds for D-π-A systems due to their relatively low resonance energies, the facile and cheap preparation of chromophores with high stabilities, and good nonlinearities [98]. The hyperpolarizabilities β of derivatives **7–9** were measured using hyper-Rayleigh scattering, and also the solvatochromic behavior of these thiophenes was investigated [99] (see **Figure 4**). A series of formyl-5-aryl-2,2′-bithiophenes **10** were synthesized by Herbivo et al. through two methods: Vilsmeier-Haack-Arnold reaction or through Suzuki coupling as precursors for NLO materials [100]. Raposo et al. synthesized two series of donor-acceptor conjugated heterocyclic azo-thienylpyrroles **11–12**. Thienylpyrroles **11** have largest first-order hyperpolarizabilities (β = 460–660 × 10^{-30} esu, T convention) [101]. Two series of novel push-pull 1-(4-(thiophen-2-yl)phenyl)-1H-pyrroles 13–14, reported by Castro et al., were studied considering the electron-accepting moieties linked at the arylthiophene bridge or to the pyrrole heterocycle, which plays the role of donor group. Thiophene ring with functionalized thiobarbituric acid derivative **14** shows the largest first hyperpolarizability (β =2480 × 10^{-30} esu) [102]. Batista et al. developed a series of thienylpyrrolyl with π-conjugated system attached to functionalized benzimidazole heterocycles **15–16** [103]. The new chromophores possess excellent solvatochromic properties and good molecular optical nonlinearities. By condensation of 5,6-phenanthroline-dione with formyl-thiophene derivatives in the presence of ammonium acetate in glacial acetic acid, oligothienyl-imidazophenanthroline compounds **17–18** were synthesized by Batista et al. These *p*-conjugated compounds possessed good values of β hyperpolarizabilities and solvatochromic properties [104]. Liu et al. synthesized isomers **19** and **20** [97], very similar in chemical structure, differing only in the substitution position of the alkyl groups attached to their conjugated cores, two novel small molecule photovoltaic materials, by implying a Knoevenagel reaction and investigated their photovoltaic properties by blending each of them with a polymer donor and a fullerene acceptor. Compounds present similar optical absorption spectra in solution and molecular energy levels in

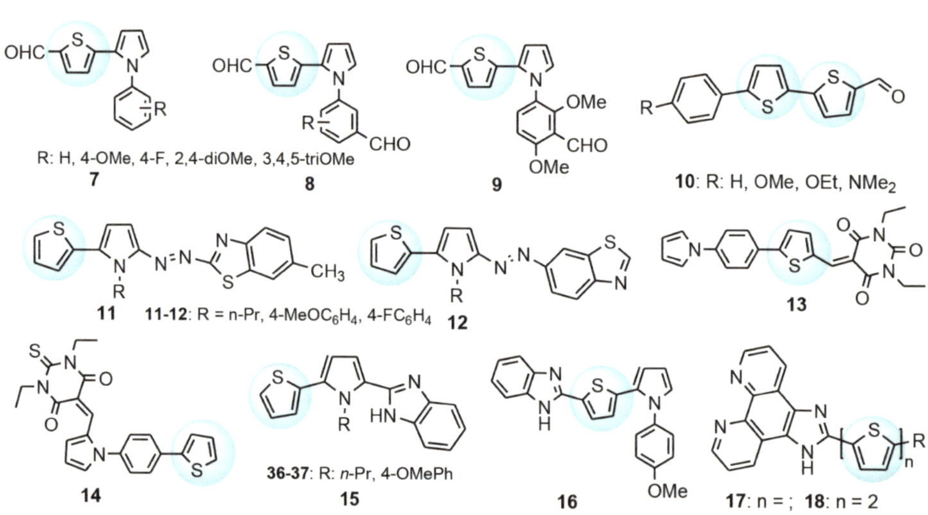

Figure 4.
Thienyl compounds with nonlinear optical properties [97–104].

a solid film. Compound **19** forms an ordered lamellar packed structure with compact π-π stacking, whereas **20** shows only a weak π-π stacking effect in solid film. Also, the authors correlate chemical structure of the A-D-A semiconductors with their aggregation structure in solid state and their electron-donating and electron-accepting properties in OPV devices (see **Figure 5**). Zhu et al. synthesized the fused-ring electron acceptors **21** and **22** with a naphtho[1,2-b,5,6-b′]dithiophene core and alkoxy side chains using a Stille reaction between an organotin compound and ethyl 2-bromothiophene-3-carboxylate [105]. It was found that change in the side chains affects electronic, optical, charge transport and morphological properties of these analogues. Compound **22** exhibits a narrower bandgap, redshifted absorption spectra, and higher electron mobility compared with **21**. The introduction of alkoxy side chains in fused-ring electron acceptors appears to be a promising approach to enhance absorption and electron mobility and therefore efficiencies in organic solar cells. Dai et al. synthesize fused-ring electron acceptors **23–24** using a Suzuki reaction and found that extending the core size leads to upshift of the energy levels, enhancement of molecular packing, higher electron mobility, more balanced-charged transport, and reduced bimolecular charge recombination (**Figure 6**).

3.4 Synthesis of the nonlinear optical dyes

Raposo et al. reported the synthesis of the 2,2′-bithiophene-conjugated dyes **25**, using the Friedel-Crafts and the Lawesson reactions. The compounds **25** have proved promising NLO chromophores (see **Figure 7**), exhibiting β_o values of 2.5–4 times higher than that of 4-nitroaniline [106]. Also, azo-dyes **26** possessed great values of molecular hyperpolarizability β_0 [107]. Marinescu et al. synthesized by azo-coupling reactions a series of pyrazole-5-ones **27–30**, which possessed only hydrazo tautomers and torsional structures, correlated with their NLO properties and the efficiency of compounds was assessed by the relationship between high β_{tot} and low HOMO-LUMO energy gap [89]. Liu et al. synthesized porphyrin dyes **31** by [2+2] click reaction, with very good yields and small energy gaps HOMO-LUMO for each compound, and demonstrated that click moieties affected the third-order nonlinear optical responses [108] (see **Figure 8**).

Pascal et al. obtained the push-pull dyes **33–34** with tricyanofuran acceptor linked to donors of various strengths, by two consecutive condensations, and investigated their second-order nonlinear optical properties [109]. It was found that a high intrinsic hyperpolarizability coupled with a high static dipolar moment results in excellent μβ values for polymethine dyes containing pyran, thiopyran,

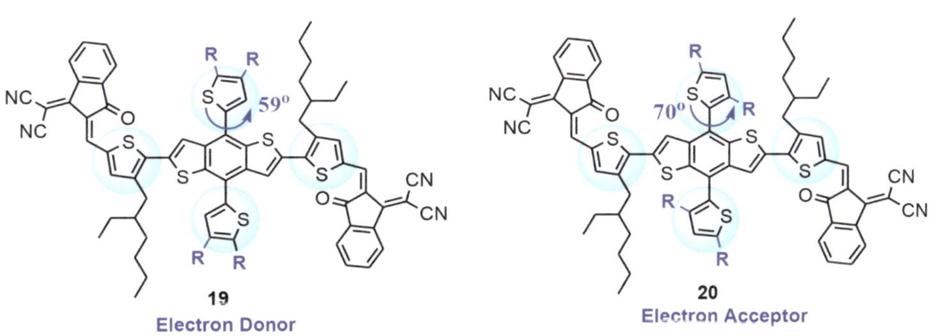

Figure 5.
Small-molecule semiconductors based on an A-D-A structure [96].

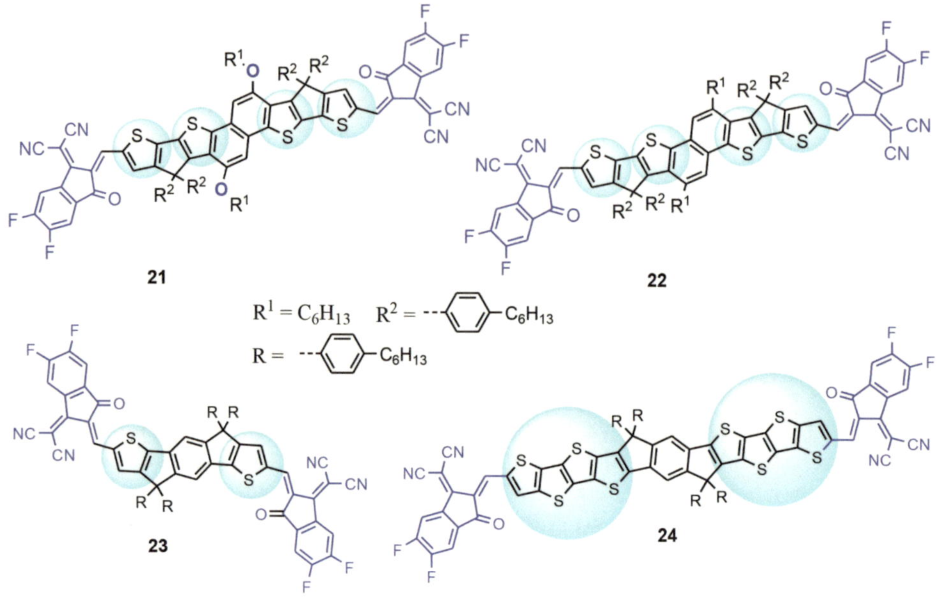

Figure 6.
Chemical structure of fused-ring electron acceptor **21–24** *[104, 105].*

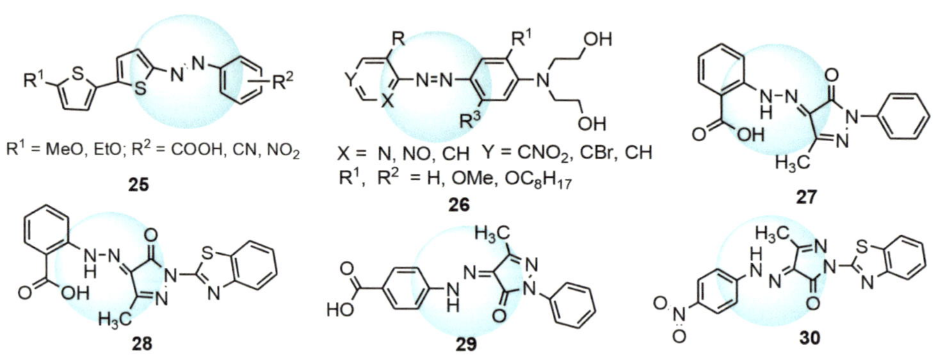

Figure 7.
Azo-dyes as NLO chromophores **25–30** *[89, 106, 107].*

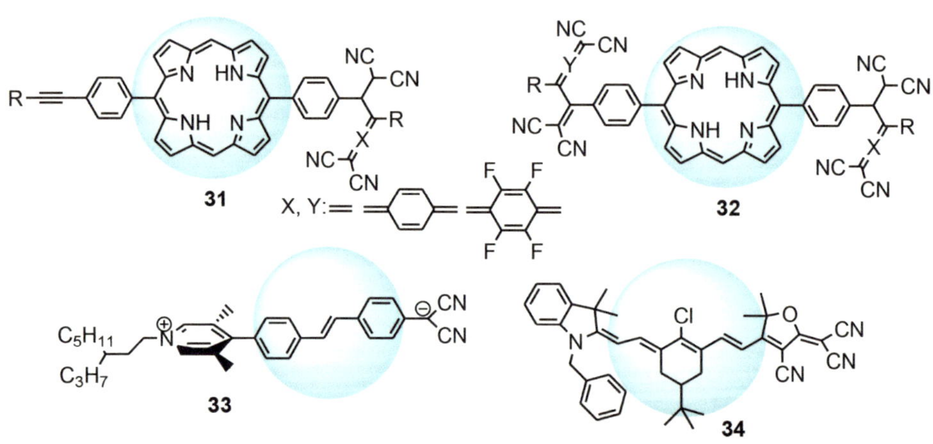

Figure 8.
Porphyrins **31–32** *and push-pull polymethine dyes* **33–34** *[108, 109].*

selenopyran, and benzoindoline moieties, and therefore, these dyes are promising for electro-optical modulation.

3.5 Synthesis of the polymers with nonlinear optical properties

A review of all the compounds deposited by MAPLE, including organics, with their applications was carried out by Caricato et al. [33]. Consequently, here we point out the newest structures reported in the literature with the most notable nonlinear optical properties mentioned.

Mariano et al. combine spin coating with the MAPLE technique, to realize polymeric multilayered thin films using three polymers **35–37** [110] (see **Figure 9**). This is the first white light-emitting diode produced by multilayer polymeric thin films reported in literature [110]. Constantinescu et al. reported fabrication of thin solid layers of poly(3,3‴-didodecyl-quaterthiophene) **38** by MAPLE, on silicon and quartz substrates and studied films with different thicknesses, morphologies, and optical properties [111]. Thin films of polyvinyl chloride **39**, polyacrylic acid **40**, and polyaniline **41** with potential use in optoelectronic were reported by Constantinescu et al. [112]. Thin films of the blue-emitting poly(9,9-dioctylfluorene) **42** deposited by MAPLE have been investigated at different laser fluence values and at different laser repetition rate. Authors reported that at high laser fluence (450 mJ/cm^2), the dioctylfluorene surface showed large bubbles presenting the intrinsic blue emission. At smooth surfaces, it can be observed that green emission becomes predominant [110].

Recent works reported new conjugated copolymers with different donor (D)-acceptor (A) motifs (see **Figure 10**) for optoelectronic devices [113]. Authors synthesized a series of four DAA copolymers **43–46** (see **Figure 11**) with thiophene donors and isoindigo-based acceptors and compared the optoelectronic properties of these four copolymers with those of the analogous. It was found that "increasing the number of acceptors in the polymer repeat unit decreased the LUMO energy of the polymer and increased the electron mobility" [113]. The authors studied the influence of acceptor length and planarity on the optoelectronic properties of donor-acceptor systems for the copolymers **43–46** with thiophene donors and isoindigo acceptors and formulated that changing the number of acceptors in the

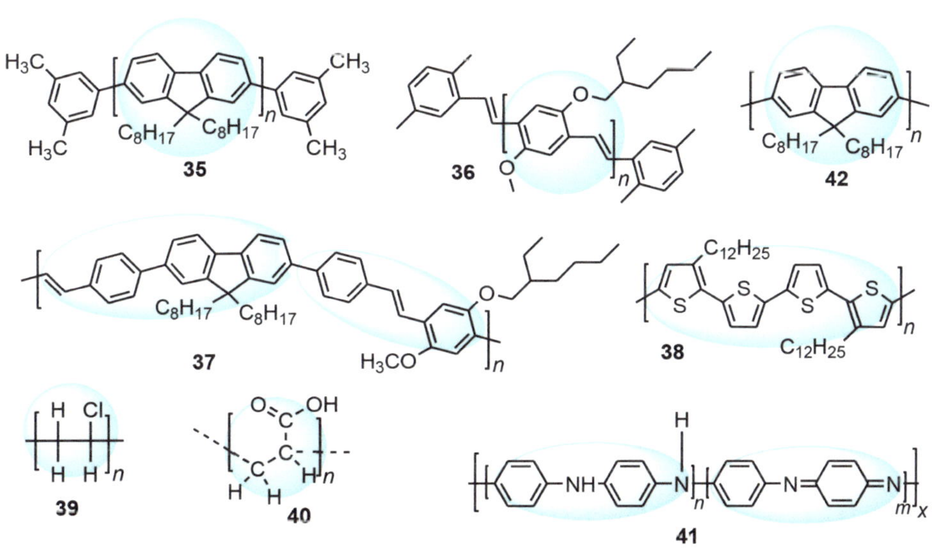

Figure 9.
Polymers deposited by MAPLE technique [110–112].

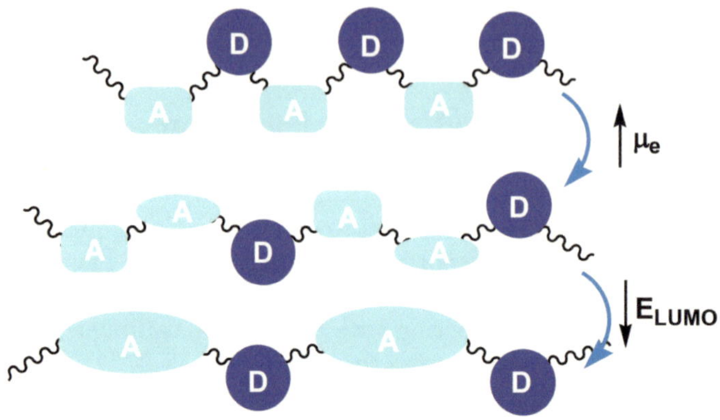

Figure 10.
Donor (D)-acceptor (A) motifs along with magnetic moment and LUMO energy variations [113].

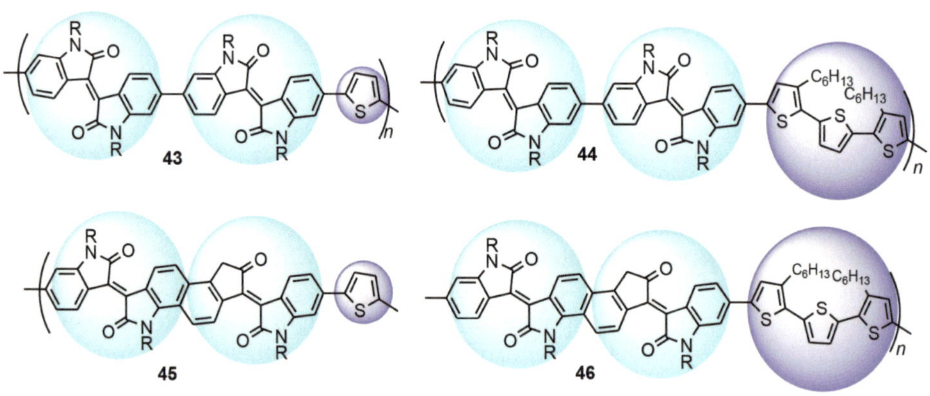

Figure 11.
Structures of DAA copolymers isoindigo-based acceptors and thiophene donors ***43–46*** *[113].*

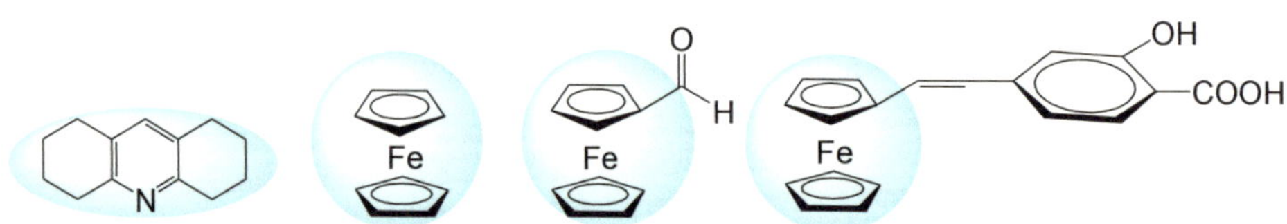

Figure 12.
Structures of compounds ***47–50*** *with NLO properties.*

polymer repeat unit is a key strategy that could be used to obtain donor-acceptor systems with fine properties and better materials.

3.6 Synthesis of other classes of organic compounds with NLO properties

Other classes of heterocyclic compounds were reported to have nonlinear optical properties, and some of them are presented below. 1,2,3,4,5,6,7,8-Octahydroacridine (OHA) **47** (see **Figure 12**) thin films grown by matrix-assisted pulsed laser evaporation (MAPLE) showed the SHG (second harmonic generation) signal of the conformational asymmetry of OHA and therefore the nonlinear optical applications of it [114]. Matei et al. studied the morphology and optical properties of ferrocene **48** thin films grown by MAPLE technique, films that were further used in two-photon absorption investigations, showing that the compound has SHG capabilities [115]. Nonlinear optical properties of ferrocene carboxaldehyde **49** thin films grown by matrix-assisted pulsed laser evaporation (MAPLE) were studied by Constantinescu et al. [116] using two-photon absorption investigations. The experiment revealed that the ferrocene **49** thin films deposited by MAPLE have second harmonic generation capabilities improved compared to simple ferrocene, with better applications in optoelectronics. 4-(Ferrocenylmethylimino)-2-hydroxy-benzoic acid 50 was synthesized by a Schiff reaction, and deposited thin films of it with controlled thickness were fabricated by matrix-assisted pulsed laser evaporation (MAPLE), on quartz and silicon substrates [117].

4. Matrix-assisted pulsed laser evaporation

The main method used to obtain thin laser films is matrix-assisted pulsed laser evaporation (MAPLE). Most organic compounds deposited by matrix-assisted pulsed laser evaporation reported so far are polymers, so they are very important for this chapter. There are three important advantages of the MAPLE technique compared to solution cast techniques: (1) the control of thickness; (2) possibility to deposit multilayers; and (3) fabrication of thin films on nonplanar substrates with good surface coverage [110].

The fact that method pulsed laser ablation is not convenient for the deposition of soft materials (almost all polymers, proteins, and other materials are chemically and/or thermally modified or destroyed) has led to the invention of a new improved method to remove these limitations. Two researchers McGill and Chrisey gave birth to matrix-assisted pulsed laser evaporation (MAPLE) technique [33] in order to deposit thin and uniform films of polymers and carbohydrates. The new method is suitable for the deposition of the complex organic materials, such as polymers, bioorganic molecules, and coordination compounds [118]. Fabrication of thin films from such materials is very important for new devices with many applications, including light-emitting diodes (LEDs) [110], field-effect transistors, sensors, photovoltaic devices, and white light sources for indoor and outdoor lighting [13, 110, 114, 115].

Three steps are necessary in the MAPLE technique:

1. dissolving the substance (solute) of interest in a volatile solvent (matrix) to form a diluted homogeneous solution (concentration of the order of 1 wt%);

2. freezing the solution at the temperature of the liquid nitrogen; and

3. placing the solution in the vacuum chamber to act as a target for laser-assisted deposition and irradiation of the frozen solution with a pulsed laser beam.

Matrix-assisted pulsed laser evaporation deposition of the desirable molecules is effectuated in a light manner, which implied the passing of the condensed phase to the gas phase. A low kinetic energy is implied in MAPLE process, in advantage to laser ablation with a high level of kinetic energy [110].

In the MAPLE method, the laser pulse energy is absorbed by the solvent and converted into thermal and kinetic energy, enabling the solvent to evaporate and carry in the gas phase the solute molecules onto the deposition substrate where they adhere as a thin film. A very volatile solvent is required to be pumped during the flight from target to substrate, and thus, the deposited film is made up of the dissolved material only.

Most of the laser energy is absorbed by the volatile matrix, not the dissolved molecules, which minimize the photochemical decomposition of the precursor solution. In addition, the use of low fluences prevents or reduces thermal damage and decomposition of molecules, so deposition can take place at low fluctuations (0.05–0.5 J/cm^2) compared to conventional pulsed laser deposition (PLD) (typically few J/cm^2).

Particularly important in this technique is the choice of solvent because it has a great impact on the deposition of organic matter, it can interact with the dissolved substance, it can lead to the production of secondary products from it, or it can be present in the deposited films [33]. The role of solvent in MAPLE technique is central; we can say that the solvent (1) must dissolve the solute without interacting with it; (2) has to be volatile; (3) must absorb laser radiation; and (4) must transport the dissolved substance from the target to the substrate.

The experimental setup of MAPLE deposition technique for thin-film fabrication is showed in **Figure 13**. The solution concentration must be of 0.1–2.0% (mass) because of the hard laser interaction with the frozen solid. The solvent is desirable to have a freezing point as high as possible. Only the solvent (also named matrix) absorbed the radiation when the laser reaches the target, so the matrix evaporates, the "solid" is ablated, and only the material's molecules are deposited on the substrate [119].

4.1 Investigation of NLO properties

4.1.1 Parameters of NLO properties

Interactions of electromagnetic fields in various media produce new fields changed in frequency, phase, amplitude, or other characteristics of the incident fields resulting nonlinear optical (NLO) properties [89]. The parameter used to evaluate the NLO susceptibility is the total hyperpolarizability (β_{tot}), meaning that a compound with large β_{tot} value is predicted to be a potential NLO active one and vice versa [11]. Literature shows that experimental determination of the β_{tot} value and therefore the NLO susceptibility is an expensive and laborious process, which led to using the quantum mechanical calculations including the DFT methods for the designing of NLO materials. The mean polarizability α, the total static dipole

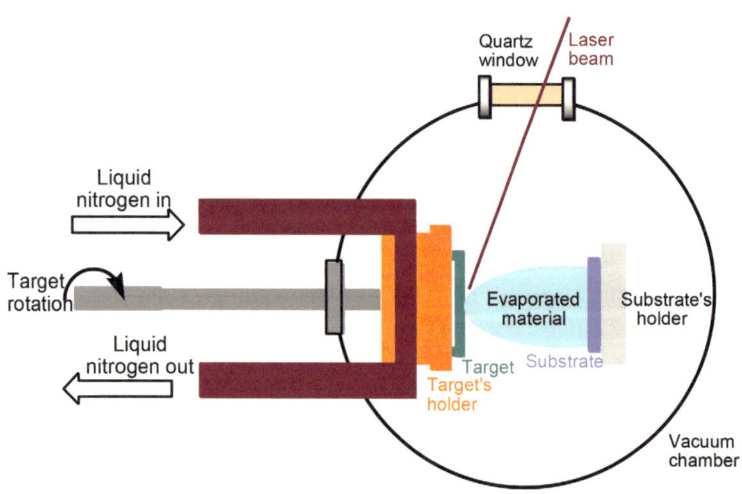

Figure 13.
Scheme of the MAPLE setup [119].

moment μ_{tot}, the quadrupole moment Q, and the mean first polarizability β_{tot} may be calculated by using DFT theory. The x, y, z components are defined as follows:

$$\alpha_{tot} = \frac{\alpha_{xx} + \alpha_{yy} + \alpha_{zz}}{3} \tag{3}$$

$$\mu_{tot} = \left(\mu_x^2 + \mu_y^2 + \mu_z^2\right)^{1/2} \tag{4}$$

$$Q_{tot} = \left(q_x^3 + q_y^3 + q_z^3\right)^{1/3} \tag{5}$$

$$\beta_{tot} = \sqrt{\left(\beta_x^2 + \beta_y^2 + \beta_z^2\right)} \tag{6}$$

$$\beta_j = \beta_{iii} + \frac{1}{3}\sum_{i \neq j}\left(\beta_{ijj} + \beta_{jij} + \beta_{jji}\right) \tag{7}$$

Therefore:

$$\beta_{tot} = \sqrt{\left(\beta_{xxx} + \beta_{xyy} + \beta_{xzz}\right)^2 + \left(\beta_{yyy} + \beta_{yzz} + \beta_{yxx}\right)^2 + \left(\beta_{zzz} + \beta_{zxx} + \beta_{zyy}\right)^2} \tag{8}$$

The larger hyperpolarizability value of one component over the other components means that the electronic charge delocalization is larger in that direction [11]. There are a lot of factors that contribute to enhance the NLO properties of the compounds, which are cumulated with the prospective β_{tot} calculated values [3].

4.1.2 Experimental determination of second harmonic generation (SHG) in thin films

Experimental setup used to investigate the SHG behavior [117] in thin-film samples is represented in **Figure 14**. The component parts of the system used for the determination of second harmonic generation (SHG) are a sapphire laser ("Tsunami," from Spectra-Physics; 780 nm, 60–100 fs pulse duration, 80 MHz repetition rate); an optical system made of a half-wave plate and a Glan-Taylor polarizing prism that allows the variation of beam intensity; a microscope's objective is to focus the laser beam onto the thin-film samples and collect the emitted SHG radiation. A dichroic mirror (DM) separated the excitation radiation, and the SHG intensity is measured by a camera spectrograph [117].

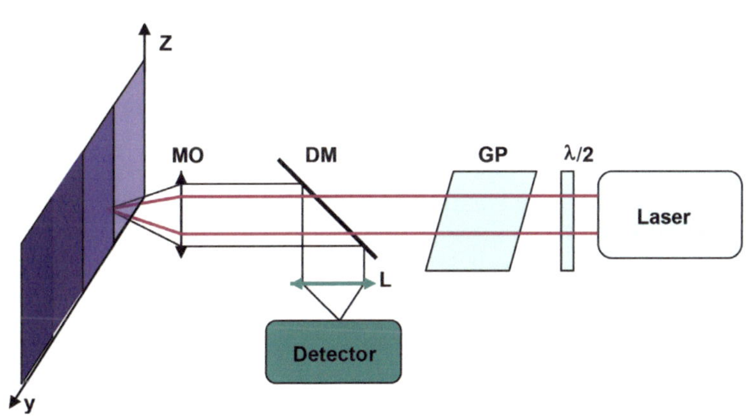

Figure 14.
Experimental setup for determining second harmonic generation.

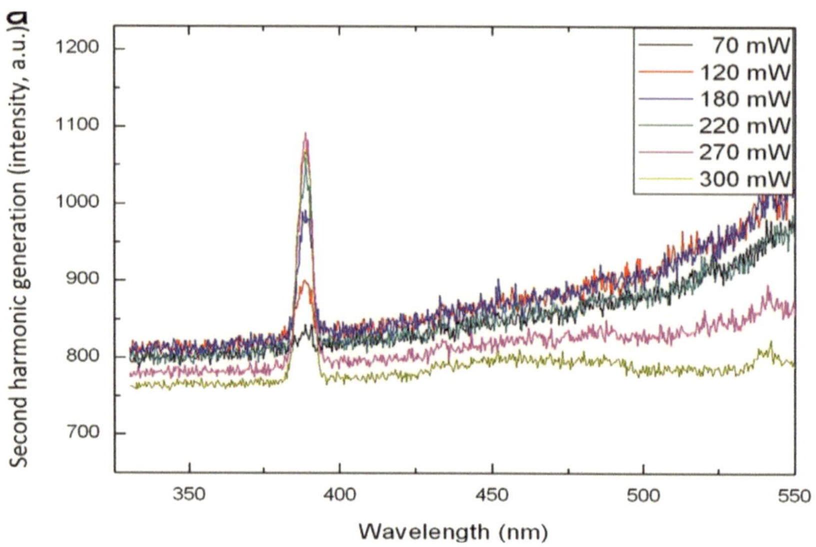

Figure 15.
The SHG experimental spectra of MAPLE-grown OHA thin films (adapted with the permission from reference [114]).

The experimental second harmonic generation of MAPLE-grown 1,2,3,4,5,6,7,8-octahydroacridine (OHA) **47** using laser fluence 0.3 J/cm^2, laser spot 2 mm^2, and 40,000 pulses, on Si substrates, under fs-laser irradiation at 788 nm wavelength is represented in **Figure 15** [114]. In this case, the presence of the SHG signal is related to the conformational asymmetry of the OHA molecule.

5. Conclusions

The synthesis of the most important classes of the nonlinear optical organic compounds, fullerenes, perylenes, thiophene, azo-dyes dyes, thienes, polymers, and other compounds, along with the techniques employed for the deposition of these compounds, was presented. For the synthesis of the new compounds with nonlinear optical applications, important reactions, like Stille, Suzuki, Knoevenagel, Huisgen, Vilsmeier-Haack-Arnold, click, were employed.

Among the simpler and more sophisticated techniques, the matrix-assisted pulsed laser evaporation (MAPLE) technique that permits making organic films with different morphologies, on different types of substrates, is the main method used to obtain thin laser films, with three basic advantages: the control of thickness; possibility to deposit multilayers; and fabrication of thin films on nonplanar substrates with good surface coverage. Crystallization growth mechanisms in MAPLE-deposited conjugated polymer films that determine specific structure, therefore the carrier transport properties, were discussed in relation with second harmonic generation (SHG) behavior of the thin films.

Organic compounds are cheap, low toxicity, ease of solution processability; therefore, their applications as NLO materials are growing.

Conflict of interest

The author has no conflict of interest to declare.

Author details

Maria Marinescu
Faculty of Chemistry, University of Bucharest, Bucharest, Romania

*Address all correspondence to: maria7marinescu@yahoo.com

References

[1] Forrest SR. Ultrathin organic films grown by organic molecular beam deposition and related techniques. Chemical Reviews. 1997;**97**:1793-1896. DOI: 10.1021/cr9410140

[2] Achelle S, Baudequin C, Ple N. Luminescent materials incorporating pyrazine or quinoxaline moieties. Dyes and Pigments. 2013;**98**:575-600. DOI: 10.1016/j.dyepig.2013.03.030

[3] Cinteza LO, Marinescu M. Synthesis and nonlinear studies on selected organic compounds in nanostructured thin films. In: Chowdhury MA, editor. Advanced Surface Engineering Research. London: IntechOpen; 2018. pp. 1-23. DOI: 10.5772/intechopen.79522

[4] Kamanina NV, Plekhanov AI. Mechanisms of optical limiting in fullerene-doped π-conjugated organic structures demonstrated with polyimide and COANP molecules. Optics and Spectroscopy. 2002;**93**:408-415

[5] Likhomanova SV, Kamanina NV. COANP-fullerenes system for optical modulation. Journal of Physics: Conference Series. 2016;**741**:012146. DOI: 10.1088/1742-6596/741/1/012146

[6] Kamanina NV, Serov SV, Shurpo NA, Likhmanova SV, Timonin DN, Kuzhakov PV, et al. Polyimide-fullerene nanostructured materials for nonlinear optics and solar energy applications. Journal of Materials Science: Materials in Electronics. 2012;**3**:1538-1542. DOI: 10.1007/s10854-012-0625-9

[7] Kamanina NV. Peculiarities of optical limiting effect in π-conjugated organic systems based on 2-cyclooctylamino-5-nitropyridinedoped with C70. Journal of Optics A: Pure and Applied Optics. 2001;**3**:321-325

[8] Denisov A, Tocnaye JLB. Soluble fullerene derivative in liquid crystal: Polymer composites and their impact on photorefractive grating efficiency and resolution. Applied Optics. 2009;**48**: 1926-1931

[9] Makowska-Janusik M. Influence of the polymeric matrix on the NLO molecular response in guest-host materials. Nonlinear Optical and Quantum Electronics. 2007;**37**:75-85

[10] Liu X, Yang Z, Wang D, Cao H. Molecular structures and second-order nonlinear optical properties of ionic organic crystal materials. Crystals. 2016; **6**:158-165. DOI: 10339/cryst6120158

[11] Wazzan N, Safi Z. DFT calculations of the tautomerization and NLO properties of 5-amino-7-(pyrrolidin-1-yl)-4-trimethyl-1,4-dihydro-1,6-naphthyridine-8-carbonitrile (APNC). Journal of Molecular Structure. 2017; **1143**:397-404. DOI: 10.1016/j.molstruct.2017.04.101

[12] Wang T, Wang X, Zhang J, Wang C, Shao J, Jiang Z, et al. Synthesis, structure and third-order optical nonlinearities of hyperbranched metal phthalocyanines containing imide units. Dyes and Pigments. 2018;**154**:75-81. DOI: 10.1016/j.dyepig.2018.02.031

[13] Pegu D, Deb J, Saha SK, Paul MK, Sarkar U. Molecular structure, chemical reactivity, nonlinear optical activity and vibrational spectroscopic studies on 6-(4-n-heptyloxybenzyoloxy)-2-hydroxy benzylidene)amino)-2H-chromen-2-one: A combined density functional theory and experimental approach. Journal of Molecular Structure. 2018; **1160**:167-176. DOI: 10.1016/j.molstruct.2018.01.090

[14] Liu J, Ykang M, Gao W, Fedorchuk AA, Kityk IV. Synthesis and nonlinear optical properties of novel conjugated small molecules based on indole donor.

Journal of Molecular Structure. 2018; **1165**:223-227. DOI: 10.1016/j.molstruc.2018.03.133

[15] Akchurin IO, Yakhutina AI, Bochkov AY, Solovjova NP, Medvedev MG, Traven VF. Novel push-pull fluorescent dyes e 7-(diethylamino) furo- and thieno[3,2-c]coumarins derivatives: Structure, electronic spectra and TD-DFT study. Journal of Molecular Structure. 2018;**1160**:215-221. DOI: 10.1016/j.molstruct.2018.01.086

[16] Ghiasuddin AM, Adeel M, Khalid M, Tahir MN, Khan MU. A combined experimental and computational study of 3-bromo-5-(2,5-difluorophenyl) pyridine and 3,5-bis(hydrate-1-yl) pyridine: Insight into the synthesis, spectroscopic, single crystal XRD, electronic, nonlinear optical and biological properties. Journal of Molecular Structure. 2018;**1160**:129-141. DOI: 10.1016/j.molstruct.2018.01.100

[17] Marinescu M, Emandi A, Marton G, Cinteza LO, Constantinescu C. Structural studies and optical nonlinear response of some pyrazole-5-ones. Nanoscience Nanotechnology Letters. 2015;7:846-854. DOI: 10. 1166/nnl.2013.1552

[18] Bragiel P, Radkowska I, Belka R, Marciniak B, Bak Z. Structural, spectroscopic and NLO features of 4-chloro-1-naphthol. Journal of Molecular Structure. 2018;**1154**:27-38. DOI: 10.1016/j.molstruc. 017.10.017

[19] Katariya SB, Patil D, Rhyman L, Alswaidan IA, Ramasami P, Sekar N. Triphenylamine-based fluorescent NLO phores with ICT characteristics: Solvatochromic and theoretical study. Journal of Molecular Structure. 2018; **1150**:493-506. DOI: 10.1016/j.molstruct.2017.08.084

[20] Drozd M, Daszkiewicz M. A synthesis, X-ray crystallographic and vibrational studies of guanidinium *o*-nitrobenzoate hydrate. New NLO crystal in guanidinium nitrobenzoate family. Journal of Molecular Structure. 2018;**1161**:383-392. DOI: 10.1016/j.molstruc. 2018.02.077

[21] Sangeetha K, Guru Prasad L, Mathammal L. Structural elucidation and physicochemical properties of an organic NLO crystal: 4-Nitrotoluene-2-sulphonic acid dihydrate. Journal of Molecular Structure. 2018;**1155**:598-609. DOI: 10.1016/j.molstruct.2017.11048

[22] Szukalski A, Sahraoui B, Kulyk B, Lazar CA, Manea AM, Mysliwiec J. Chemical structure *versus* second-order nonlinear optical response of the push–pull type pyrazoline-based chromophores. RSC Advances. 2017; **2017**:9941-9947. DOI: 10.1039/C6RA26781E

[23] Zhang YX, Wang YH. Nonlinear optical properties of metal nanoparticles: A review. RSC Advances. 2017;**2017**:45129-45144. DOI: 10.039/C7RA0755K

[24] Han P, Wang D, Gao H, Zhang J, Xing Y, Yang Z, et al. Third-order nonlinear optical properties of cyanine dyes with click chemistry modification. Dyes and Pigments. 2018;**149**:8-15. DOI: 10.1016/j.dyepig.2017.09.052

[25] Medishetty R, Zareba JK, Mayer D, Samoc M, Fischer RA. Nonlinear optical properties, upconversion and lasing in metal–organic frameworks. Chemical Society Reviews. 2017;**46**:4976-5004. DOI: 10.1039/c7cs00162b

[26] Fernandes SSM, Herbivo C, de-Sousa J, Comel A, Belsley M, Raposo MMM. Theoretical and experimental studies of aryl-bithiophene based push-pull π-conjugated heterocyclic systems bearing cyanoacetic or rhodanine-3-acetic acid acceptors for SHG NLO applications. Dyes and Pigments. 2018; **149**:566-573. DOI: 10.1016/j.dyepig.2017.10.00 1

[27] Čibová A, Martinická A, Magdolen P, Zahradník P, Cigáň M, Uherek M, et al. Dicationic and monocationic benzobisthiazolium salts as potential NLO chromophores. Dyes and Pigments. 2018;**149**:597-611. DOI: 10.1016/j.dyepig.2017.11.015

[28] Xu H, Wan N, Zhang X, Li Z, Deng G. Monolithic NLO chromophores with different pendant groups and matrix-assisted-poling effect: Synthesis and characterization. Dyes and Pigments. 2018;**157**:230-237. DOI: 10.1016/j.dyepig.2018.04.061

[29] Fominykh OD, Kalinin AA, Sharipova SM, Sharipova AV, Burganov TI, Smirnov MA, et al. Composite materials containing chromophores with 3,7-(di)vinylquinoxalinone π-electron bridge doped into PMMA: Atomistic modeling and measurements of quadratic nonlinear optical activity. Dyes and Pigments. 2018;**158**:131-141. DOI: 10.1016/j.Dyepig.2018.05.033

[30] Durand RJ, Achelle S, Gauthier S, Cabon N, Ducamp M, Kahlal S. Incorporation of a ferrocene unit in the π-conjugated structure of donor-linker-acceptor (D-π-A) chromophores for nonlinear optics (NLO). Dyes and Pigments. 2018;**155**:68-74. DOI: 10.1016/j.dyepig.2018.03.029

[31] Erande Y, Kothavale S, Sreenath MC, Chitrambalam S, Joe IH, Sekar N. Triphenylamine derived coumarin chalcones and their red emitting OBO difluoride complexes: Synthesis, photophysical and NLO property study. Dyes and Pigments. 2018;**148**:474-491. DOI: 10.1016/j.dyepig.2017. 09.045

[32] Nie W. Optical nonlinearity: Phenomena, applications, and materials. Advanced Materials. 1993;**5**:520-545. DOI: 10.1002/adma.19930050704

[33] Caricato AP, Ge W, Stiff-Roberts AD. UV- and RIR-MAPLE: Fundamentals and applications. In: Ossi PM, editor. Advances in the Application of Lasers in Materials Science. Springer; 2018. pp. 275-308. DOI: 10.1007/978-3-319-96845-2_10

[34] Mukherjee S, Thilagar P. Organic white-light emitting materials. Review. Dyes and Pigments. 2014;**110**:2-27. DOI: 10.1016/j.dyepig.2014.05.031

[35] Chopra KL, Paulson PD, Dutta V. Thin-film solar cells: An overview. Progress in Photovoltaics: Research and Applications. 2004;**12**:69-92. DOI: 10.1002/pip.541

[36] Baxamusa S. Conformal polymer CVD (chapter 5). In: Gleason KK, editor. CVD Polymers: Fabrication of Organic Surfaces and Devices. Wiley; 2015. pp. 87-109. DOI: 10.1002/9783527690275.ch5

[37] Kumar DU, Kiran AJ, Murali MG, Adhikari AV. Donor-acceptor conjugated polymers and their nanocomposites for photonic applications (chapter 6). In: Kamanina N, editor. Nonlinear Optics. IntechOpen; 2012. pp. 133-160. DOI: 10.5772/2073

[38] Lee YL, Chen CH, Yang YM. Surface morphology and wetting behavior of poly(α-methylstyrene) thin films prepared by vacuum deposition. Langmuir. 1998;**14**:6980-6986. DOI: 10.1021/la9800607

[39] Tang Y, Wang Y, Wang G, Wang H, Wang L, Yan D. Vacuum-deposited submonolayer thin films of a three-ring bent-core compound. Journal of Physical Chemistry B. 2004;**108**:12921-12926. DOI: 10.1021/jp048607f

[40] Volpati D, Machado AD, Olivati CA, Alves N, Curvelo AAS, Pasquini D, et al. Physical vapor deposited thin films of lignins extracted from sugar cane bagasse: Morphology, electrical properties, and sensing applications.

Biomacromolecules. 2011;**12**:3223-3231. DOI: 10.1021/bm200704m

[41] Costa JCS, Rocah RM, Vaz ICM, Torres MC, Mendes A, Santos LMNBF. Description and test of a new multilayer thin film vapor deposition apparatus for organic semiconductor materials. Journal of Chemical Engineering Data. 2015;**60**:3776-3791. DOI: 10.1021/acs.jced.5b00708

[42] Esaki Y, Komio T, Matsushima T, Adachi C. Enhanced electrical properties and air stability of amorphous organic thin films by engineering film density. The Journal of Physical Chemisty Letters. 2017;**8**: 5891-5897. DOI: 10.1021/acs.jpclett.7b02808

[43] Baldo M, Deutsch M, Burrows P, Gossenberger H, Gerstenberg M, Ban V, et al. Organic vapor phase deposition. Advanced Materials. 1998;**10**:1505-1514. DOI: 10.1002/(SICI)1521-4095(199812)10:18 <1505::AID-ADMA1505>3.0.CO;2-G

[44] Fraxedas J. Perspectives of thin molecular organic films. Advanced Materials. 2002;**14**:1603-1614. DOI: 10.1002/1521-4095(20021118) 14:22<1603::AID-ADMA1603>3.0.CO;2-5

[45] Lunt RR, Benziger JB, Forrest SR. Growth of an ordered crystalline organic heterojunction. Advanced Materials. 2007;**19**:4229-4233. DOI: 10.1002/adma.200701572

[46] Wang H, Zhu F, Yang J, Geng Y, Yan D. Weak epitaxy growth affording high-mobility thin films of disk-like organic semiconductors. Advanced Materials. 2007;**19**:2168-2171. DOI: 10.1002/adma.200602566

[47] Witte G, Woll C. Molecular beam deposition and characterization of thin organic films on metals for applications in organic electronics. Physica Status Solidi A. 2008;**205**:497-510. DOI: 10.1002/pssa.200723433

[48] Hagen JA, Li WX, Spaeth H, Grote JG, Steckl AJ. Molecular beam deposition of DNA nanometer films. Nano Letters. 2007;**7**:133-137. DOI: 10.1021/nl062342u

[49] Yamashita A, Tayashi T. Organic molecular beam deposition of metallophthalocyanines for opto-electronics applications. Advanced Materials. 1996;**8**:791-799. DOI: 10.1002/adma.19960081004

[50] Borghesi A, Sassella A, Tubino R, Destri S, Porzio W. Organic molecular beam deposition of highly oriented β-tetrahexylsexithiophene films. Advanced Materials. 1998;**10**:931-934. DOI: 10.1002/(SICI)1521-4095(199808)10:12<931::AID-ADMA931>3.0.CO;2-#

[51] Bayliss SM, Heutz S, Clots R, Middleton RL, Rumbles G, Jones TS. Templating effects in the growth of metal-free phthalocyanine polymorphic double layers. Advanced Materials. 2000;**12**:202-206. DOI: 10.1002/(SICI)1521-4095(200002)12:3<202::AID-ADMA202>3.0

[52] Chao H, Koski J, Riggleman RA. Solvent vapor annealing in block copolymer nanocomposite films: A dynamic mean field approach. Soft Matter. 2017;**13**:239-249. DOI: 10.1039/c6sm00770h

[53] Wessendorf CD, Perez-Rodriguez A, Hanisch J, Arndt A, Ata I, Schulz GL, et al. Understanding the effect of solvent vapor annealing on solution-processed A-D-A oligothiophene bulk-heterojunction solar cells: The role of alkyl side chains. Journal of Materials Chemistry A. 2016;**4**:2571-2580. DOI: 10.1039/C5TA07713C

[54] Kim K, Park S, Kim Y, Ban J, Park C, Ryu DY. Optimized solvent vapor annealing for long-range perpendicular

lamellae in PS-b-PMMA films. Macromolecules. 2016;**49**:1722-1730. DOI: 10.1021/acs.macromol. 5b02188

[55] Nelson G, Drapes CS, Grant MA, Gnabasik R, Wong J, Baruth A. High-precision solvent vapor annealing for block copolymer thin films. Micromachines. 2018;**9**:271. DOI: 10.3390/mi9060271

[56] Tomita K, Shioya N, Kise R, Shimoaka T, Yoshida H, Koganezawad T, et al. Structure control of a zinc tetraphenylporphyrin thin film by vapor annealing using fluorine containing solvent. Thin Solid Films. 2018;**665**: 85-90. DOI: 10.1016/j.tsf.2018.09.015

[57] Heflin JR, Guzy MT, Neyman PJ, Gaskins KJ, Brands C, Wang Z, et al. Efficient, thermally stable, second order nonlinear optical response in organic hybrid covalent/ionic self-assembled films. Langmuir. 2006;**22**:5723-5727. DOI: 10.1021/la0602970

[58] Kroetch A, Buswell SC, Evoy S. Fabrication of nonlinear optical devices in ionically self-assembled monolayers. Journal of Micro/Nanolithography, MEMS, and MOEMS. 2009;**8**:013011. DOI: 10.1117/1.3066521

[59] Teichler A, Perelaer J, Schubert US. Inkjet printing of organic electronics comparison of deposition techniques and state-of-the-art developments. Journal of Materials Chemistry C. 2013; **1**:1910-1925. DOI: 10.1039/ C2TC00255H

[60] Amruth C, Luszczynska B, Dupont BGR, Sieradzki Z. Inkjet printing technique and its application in organic light emitting diodes. Display and Imaging. 2017;(2):339-358. DOI: 10.1117/2.1201506.005940

[61] Ringeisen BR, Callahan J, Wu PK, Pique A, Spargo B, RA MG, et al. Novel laser-based deposition of active protein thin films. Langmuir. 2001;**17**: 3472-3479. DOI: 10.1021/la0016874

[62] Ouazzani HE, Dabos-Seignon S, Gindre D, Iliopoulos K, Todorova M, Bakalska R, et al. Novel styrylquinolinium dye thin films deposited by pulsed laser deposition for nonlinear optical applications. The Journal of Physical Chemistry C. 2012; **116**:7144-7152. DOI: 10.1021/jp2118218

[63] Majumdar S, Grochowska K, Sawczak M, Śliwiński G, Huhtinen H, Dahl J, et al. Interfacial properties of organic semiconductor–inorganic magnetic oxide hybrid spintronic systems fabricated using pulsed laser deposition. ACS Applied Materials Interfaces. 2015;**7**:22228-22237. DOI: 10.1021/acsami.5b04840

[64] Fischer D, Mankowski A, Ranft A, Vasa SK, Linser R, Mannhart J, et al. ZIF-8 films prepared by femtosecond pulsed-laser deposition. Chemistry of Materials. 2017;**29**:5148-5155. DOI: 10.1021/acs.chemmater. 7b00611

[65] Sima F, Mutlu EC, Eroglu MS, Sima LE, Serban N, Ristoscu C, et al. Levan nanostructured thin films by MAPLE assembling. Biomacromolecules. 2011; **12**:2251-2256. DOI: 10.1021/bm200340b

[66] Jeong H, Napolitano S, Arnold CB, Priestley RD. Irreversible adsorption controls crystallization in vapor-deposited polymer thin films. Journal of Physical Chemistry Letters. 2017;**8**: 229-234. DOI: 10.1021/acs.jpclett. 6b02573

[67] Jeong H, Chowdhury M, Wang Y, Sezen-Edmonds M, Loo YL, Register RA, et al. Tuning morphology and melting temperature in polyethylene films by MAPLE. Macromolecules. 2018;**51**:512-519. DOI: 10.1021/acs. macromol.7b02345

[68] Liu Y, Hu W, Xu Y, Liu S, Zhu D. Characterization, second-harmonic

generation, and gas-sensitive properties of Langmuir-Blodgett films of 1,8-naphthalimide-tri-tert-butylphthalocyanine. Langmuir. 2000; **104**:11859-11863. DOI: 10.1021/jp0018798

[69] Modlinska A, Filipowicz M, Martynski T. Bulk organisation and alignment in Langmuir and Langmuir-Blodgett films of tetrachloroperylene tetracarboxylic acid esters. Journal of Molecular Structure. 2016;**1125**:680-687. DOI: 10.1016/j.molstruc.2016.07.054

[70] Ronchi M, Biroli AO, Marinotto D, Pizzotti M, Ubaldi MC, Pietralunga SM. The role of the chromophore size and shape on the SHG stability of PMMA films with embebbed NLO active macrocyclic chromophores based on a cyclotetrasiloxane scaffold. Journal of Physical Chemistry C. 2011;**115**: 4240-4246. DOI: 10.1021/jp1107476

[71] Dragonetti C, Colombo A, Fontani M, Marinotto D, Nisic F, Righetto S, et al. Novel fullerene platinum alkynyl complexes with high second-order nonlinear optical properties as a springboard for NLO-active polymer films. Organometallics. 2016;**35**: 1015-1021. DOI: 10.1021/acs.organomet.6b00094

[72] Brinker CJ. Dip coating (chapter 10). In: Schneller T, Waser R, Kosec M, Payne D, editors. Chemical Solution Deposition of Functional Oxide Thin Films. Wien: Springer-Verlag; 2013. pp. 233-261. DOI: 10.1007/978-3-211-99311-8_10

[73] Huang L, Hu X, Chi L. Monolayer-mediated growth of organic semiconductor films with improved device performance. Langmuir. 2015;**31**: 9748-9761. DOI: 10.1021/acs.langmuir.5b00349

[74] Ohishi T. Preparation and properties of sol–gel thin film containing rhodamine B derivative with ethoxy silano group on organic substrate and its application to surface-treatment thin film for display. Journal of Sol-Gel Science and Technology. 2004;**32**: 281-285. DOI: 10.1007/s10971-004-5803-7

[75] Kang EH, Bu T, Jin P, Sun J, Yang Y, Shen J. Layer-by-layer deposited organic/inorganic hybrid multilayer films containing noncentro symmetrically orientated azobenzene chromophores. Langmuir. 2007;**23**: 7594-7601. DOI: 10.1021/la700749s CCC

[76] Perednis D, Gauckler LJ. Thin film deposition using spray pyrolysis. Journal of Electroceramics. 2005;**14**:103-111. DOI: 10.1007/s10832-005-0870-x

[77] Karaman M, Cabuk N. Initiated chemical vapor deposition of pH responsive poly(2-diisopropylamino) ethyl methacrylate thin films. Thin Solid Films. 2012;**520**:6484-6488. DOI: 10.1016/j.tsf.2012.06.083

[78] Jeevendrakumar VJB, Altemus BA, Gildea AJ, Bergkvist M. Thermal properties of poly (neopentylmethacrylate) thin films deposited via solventless, radical initiated chemical vapor deposition. Thin Solid Films. 2013;**542**:81-86. DOI: 10.1016/j.tsf.2013.06.056

[79] Christian P, Coclite AM. Thermal studies on proton conductive copolymer thin films based on perfluoroacrylates synthesized by initiated chemical vapor deposition. Thin Solid Films. 2017;**63**: 3-8. DOI: 10.1016/j.tsf.2017.01.023

[80] Coskun H, Aljabour A, Uiberlacker L, Strobel M, Hild S, Cobet C, et al. Chemical vapor deposition–based synthesis of conductive polydopamine thin-films. Thin Solid Films. 2018;**645**: 320-325. DOI: 10.1016/j.tsf.2017.10.063

[81] Chen WY, Matthews A, Jones FR, Chen KS. Immobilization of carboxylic

acid groups on polymeric substrates by plasma-enhanced chemical vapor or atmospheric pressure plasma deposition of acetic acid. Thin Solid Films. 2018; **666**:54-60. DOI: 10.1016 /j. tsf.2018.07.051

[82] Ozpirin M, Ebil O. Transparent block copolymer thin films for protection of optical elements via chemical vapor deposition. Thin Solid Films. 2018;**660**:391-398. DOI: 10.1016/ j.tsf.2018.06.044

[83] McAfee SM, Dayneko SV, Josse P, Blanchard P, Cabanetos C, Welch GC. Simply complex: The efficient synthesis of an intricate molecular acceptor for high-performance air-processed and air-tested fullerene-free organic solar cells. Chemistry of Materials. 2017;**29**: 1309-1314. DOI: 10.1021/acs. chemmater.6b04862

[84] Cao FY, Huang WC, Chang SL, Cheng YJ. Angular-shaped 4,9-dialkylnaphthodithiophene-based octacyclic ladder-type non-fullerene acceptors for high efficiency ternary-blend organic photovoltaics. Chemistry of Materials. 2018;**30**:4968-4977. DOI: 10.1021/acs.chem mater.8b01089

[85] Matthews R, Swisher J, Hutchins KM, Pentzer EB. Perylene diimide bearing different trialkyl silyl ethers: Impact of asymmetric functionalization on self-assembly into nanostructures. Chemistry of Materials. 2018;**30**: 3571-3577. DOI: 10.1021/acs. chemmater.8b01543

[86] Wang F, Dai Y, Wang W, Lu H, Qiu L, Ding Y, et al. Incorporation of heteroatoms in conjugated polymers backbone toward air-stable, high-performance n-channel unencapsulated polymer transistors. Chemistry of Materials. 2018;**30**:5451-5459. DOI: 10.1021/acs.chemmater. 8b02359

[87] Karabacak M, Karaca C, Atac A, Eskici M, Karanfil A, Kose E. Synthesis, analysis of spectroscopic and nonlinear optical properties of the novel compound: (S)-N-benzyl-1-phenyl-5-(thiophen-3-yl)-4-pentyn-2-amine. Spectrochimica Acta A. 2012;**97**:556-567. DOI: 10.1016/j.saa.2012. 05.087\

[88] Li M, Li Y, Zhang H, Wang S, Ao Y, Cui Z. Molecular engineering of organic chromophores and polymers for enhanced bulk second-order optical nonlinearity. Journal of Materials Chemistry C. 2017;**5**:4011-4122. DOI: 10.1039/C7TC00713B

[89] Marinescu M, Emandi A, Marton G, Cinteza LO, Constantinescu C. Structural studies and optical nonlinear response of some pyrazole-5-ones. Nanoscience and Nanotechnology Letters. 2015;7:1-9. DOI: 10.1166/ nnl.2015.2032

[90] Jeong MY, Cho BR. Octupolar molecules for nonlinear optics: From molecular design to crystals and films with large second-harmonic generation. The Chemical Record. 2014;**15**:132-142. DOI: 10.1002/tcr.201402013

[91] Canulescu S, Schou J, Nielsen SF. Processing of C60 thin films by matrix-assisted pulsed laser evaporation (MAPLE). Applied Physics A. 2011;**104**: 775-780. DOI: 10.1007/s00339-011-6432-y

[92] Labrunie A, Gorenflot J, Babics M, Alévêque O, Dabos-Seignon S, Balawi AH, et al. Triphenylamine-based push–pull σ–C60 dyad as photoactive molecular material for single-component organic solar cells: Synthesis, characterizations, and photophysical properties. Chemistry of Materials. 2018;**30**:3474-3485. DOI: 10.1021/acs.chemmater.8b01117

[93] Kim Y, Choi J, Lee C, Kim Y, Kim C, Nguyen TL, et al. Aqueous soluble fullerene acceptors for efficient eco-friendly polymer solar cells processed from benign ethanol/water mixtures.

Chemistry of Materials. 2018;**30**: 5663-5672. DOI: 10.1021/acs. chemmater.8b02086

[94] Kamanina V. Introductory chapter: Why the study of fullerenes is so important? In: Kamanina V, editor. Fullerenes and Relative Materials. London: IntechOpen; 2018. pp. 1-6. DOI: 10.5772/intehopen.74812

[95] Dai S, Xiao Z, Xue P, Rech JJ, Liu K, Li Z, et al. Effect of core size on performance of fused-ring electron acceptors. Chemistry of Materials. 2018; **30**:5390-5396. DOI: 10.1021/acs. chemmater.8b02222

[96] Carlotti B, Cai Z, Kim H, Sharapov V, Madu IK, Zhao D, et al. Charge transfer and aggregation effects on the performance of planar vs twisted nonfullerene acceptor isomers for organic solar cells. Chemistry of Materials. 2018;**30**:4263-4276. DOI: 10.1021/acs.chemmater.8b01047

[97] Liu D, Yang L, Wu Y, Wang X, Zeng Y, Han G, et al. Tunable electron donating and accepting properties achieved by modulating the steric hindrance of side chains in A-D-A small-molecule photovoltaic materials. Chemistry of Materials. 2018;**30**: 619-628. DOI: 10.1021/acs. chemmater.7b03142

[98] Marco AB, Andreu R, Franco S, Garín J, Orduna J, Villacampa B, et al. Efficient second-order nonlinear optical chromophores based on dithienothiophene and thienothiophene bridges. Tetrahedron. 2013;**69**: 3919-3926. DOI: 10.1016/j. tet.2013.03.027

[99] Batista RMF, Costa SPG, Malheiro EL, Belsley M, Raposo MMM. Synthesis and characterization of new thienylpyrrolyl-benzothiazoles as efficient and thermally stable nonlinear optical chromophores. Tetrahedron. 2007;**63**:4258-4265

[100] Herbivo C, Comel A, Kirsch G, Raposo MMM. Synthesis of 5-aryl-5′-formyl-2,2′-bithiophenes as new precursors for nonlinear optical (NLO) materials. Tetrahedron. 2009;**65**: 2079-2086. DOI: 10.1016/j.tet. 2008.12. 078

[101] Raposo MMM, Castro MCR, Fonseca AMC, Schellenberg P, Belsley M. Design, synthesis and characterization of the electrochemical, nonlinear optical properties, and theoretical studies of novel thienylpyrrole azo dyes bearing benzothiazole acceptor groups. Tetrahedron. 2011;**67**:5189-5198. DOI: 10.1016/j.tet.2011.05.053

[102] Castro MCR, Belsley M, Fonseca AMC, Raposo MMM. Synthesis and characterization of novel second-order NLO-chromophores bearing pyrrole as an electron donor group. Tetrahedron. 2012;**68**:8147-8155. DOI: 10.1016/j. tet.2012.07.082

[103] Batista RMF, Costa SPG, Belsley M, Raposo MMM. Synthesis and second-order nonlinear optical properties of new chromophores containing benzimidazole, thiophene, and pyrrole heterocycles. Tetrahedron. 2007;**63**: 9842-9849. DOI: 10.1016/j. tet.2007.06.098

[104] Batista RMF, Costa SPG, Belsley M, Lodeiro C, Raposo MMM. Synthesis and characterization of novel (oligo)thienyl-imidazo-phenanthrolines as versatile π-conjugated systems for several optical applications. Tetrahedron. 2008;**64**:9230-9238. DOI: 10.1016/j.tet.2008.07.043

[105] Zhu J, Xiao Y, Wang J, Liu K, Jiang H, Lin Y, et al. Alkoxy-induced near-infrared sensitive electron acceptor for high-performance organic solar cells. Chemistry of Materials. 2018;**30**: 4150-4156. DOI: 10.1021/acs. chemmater.8b01677

[106] Raposo MMM, Ferreira AMFP, Belsley M, JCVP M. 5′-Alkoxy-2,2′-bithiophene azo dyes: A novel promising series of NLO-chromophores. Tetrahedron. 2008;**64**:5878-5884. DOI: 10.1016/j.tet.2008.04.061

[107] Kleinpeter E, Koch A, Mikhova B, Stamboliyska BA, Kolev TM. Quantification of the push–pull character of the isophorone chromophore as a measure of molecular hyperpolarizability for NLO applications. Tetrahedron Letters. 2008; **49**:1323-1327. DOI: 10.1016/j.tet.2007.11.053

[108] Liu X, Wang D, Gao H, Yang Z, Xing Y, Cao H, et al. Nonlinear optical properties of symmetrical and asymmetrical porphyrin derivatives with click chemistry modification. Dyes and Pigments. 2016;**134**:155-163. DOI: 10.1016/j.dyepig.2016.07.010

[109] Pascal S, Getmanenko YA, Zhang Y, Davydenko I, Ngo MH, Pilet G, et al. Design of near-infrared-absorbing unsymmetrical polymethine dyes with large quadratic hyperpolarizabilities. Chemistry of Materials. 2018;**30**: 3410-3418. DOI: 10.1021/acs.chemmater. 8b00960

[110] Mariano F, Caricato AP, Accorsi G, Leo C, Cesaria M, Carallo S, et al. White multi-layered polymer light emitting diode by matrix assisted pulsed laser evaporation. Journal of Materials Chemistry C. 2016;**4**:7667-7674. DOI: 10.1039/C6TC01826B

[111] Constantinescu C, Rapp L, Rotaru P, Delaporte P, Alloncle AP. Pulsed laser processing of poly(3,3″-didodecyl quarter thiophene) semiconductor for organic thin film transistors. Chemical Physics. 2015;**450-451**:32-38. DOI: 10.1016/j.chemphys.2015.02.004

[112] Constantinescu C, Rotaru P, Nedelcea A, Dinescu M. Thermal behavior and matrix-assisted pulsed laser evaporation deposition of functional polymeric materials thin films with potential use in optoelectronics. Materials Science in Semiconductor Processing. 2015;**30**: 242-249. DOI: 10.1016/j.mssp.2014.10.011

[113] Randell NM, Radford CL, Yang J, Quinn J, Hou D, Li Y, et al. Effect of acceptor unit length and planarity on the optoelectronic properties of isoindigo–thiophene donor–acceptor polymers. Chemistry of Materials. 2018; **30**:4864-4873. DOI: 10.1021/acs.chemmater.8b02535

[114] Ion V, Matei A, Constantinescu C, Ionita I, Marinescu M, Dinescu M, et al. Octahydroacridine thin films grown by matrix-assisted pulsed laser evaporation for NLO applications. Material Science in Semiconductor Processing. 2015;**36**: 78-83. DOI: 10.1016/j.mssp.2015.02.064

[115] Matei A, Constantinescu C, Ion V, Mitu B, Ionita I, Dinescu M, et al. Ferrocene, an old molecule with bright future: Thin films grown by matrix-assisted pulsed laser evaporation for nonlinear optical applications. The Journal of Organic Chemistry. 2014;**751**: 638-643. DOI: 10.1016/j.jorganchem.2013.10.056

[116] Constantinescu C, Matei A, Ion V, Mitu B, Ionita I, Dinescu M, et al. Ferrocene carboxaldehyde thin films grown by matrix-assisted pulsed laser evaporation for nonlinear optical applications. Applied Surface Science. 2014;**302**:83-86. DOI: 10.1016/j.mssp.2015.02.064

[117] Matei A, Marinescu M, Constantinescu C, Ion V, Mitu B, Ionita I, et al. Nonlinear optical studies on 4-(ferrocenylmethylimino)-2-hydroxy-benzoic acid thin films deposited by MAPLE. Applied Surface Science. 2016; **374**:206-212. DOI: 10.1016/j.apsusc.2015.11.024

[118] Chrisey DB, Pique A, McGill RA, Horwitz JS, Ringeisen BR. Laser deposition of polymer and biomaterial films. Chemistry Review. 2003;**103**: 553-576. DOI: 10.1021/cr010428w

[119] Rotaru A, Constantinescu C, Rotaru P, Moanta A, Dumitru M, Socaciu M, et al. Thermal analysis and thin film deposition by matrix assisted pulsed laser evaporation of a 4CN type azomonoether. Journal of Thermal Analysis and Calorimetry. 2008;**92**: 279-284. DOI: 10.1007/s10973-007-8818-9

Chapter 2

Magnetic Bio-Derivatives: Preparation and Their Uses in Biotechnology

Mariana Paola Cabrera, David Fernando Morais Neri, Fernando Soria and Luiz Bezerra Carvalho Jr

Abstract

In this chapter, the authors will show different proposals of magnetic bio-derivatives and its applicability in biotechnology. The historical context of immobilized enzymes, as well as highlighting the main advantages and disadvantages of each, will be mentioned. Besides, iron oxides and composite materials will be presented as support for biomolecules immobilization. Composites are effortlessly prepared including many materials capable of providing advantages to the magnetic derivatives. Enzymes covalently linked to these magnetic particles combine their catalytic properties with reaction specificity, reusability, and possible reactor construction. In addition, proteins can also be purified by these magnetic composites containing specific ligands allowing reactors and reuses too. Some characterization techniques used to study the magnetic material and derivative immobilized will be described as well. Altogether, an engaging presentation about the interesting features of magnetic bio-derivatives will highlight their uses in biotechnology field as well as in others.

Keywords: iron oxides, composite, enzyme, protein, immobilization, purification

1. Introduction

Even today, the widespread use of magnetic particles in areas such as biotechnology, engineering, material sciences, biomedicine, and microbiology, among others, still is disclosed in the literature. So, a large part of the scientific community including biomedical, biologists, and pharmacists, in addition to chemists and physicists, is looking for novel applications for magnetic bio-derivatives obtained from biomolecules immobilized on iron oxide particles. The use of magnetic particles is further than matrices for biomolecules immobilization such particles can also be used as a potential contrast agent in magnetic resonance imaging (MRI), drug delivery, magnetic hyperthermia, photothermal therapy, and food treatment in order to better the organoleptic properties [1].

The great importance of developing attractive and promising magnetic bio-derivatives as well as bringing to light new applications is evident. Our research group has proposed a considerable number of magnetic materials from inorganic (e.g., diatomaceous earth) or organic (e.g., azocasein) compounds and iron oxides as support to enzyme immobilization or protein purification. Recently, magnetic diatomaceous earth coated with polyaniline (mDE@PANI) exhibited good features as a matrix for covalent immobilization of three industrial enzymes (invertase, trypsin, and β-galactosidase) [2]. This magnetic composite was also promising for the treatment of boldo tea after immobilization of the enzyme tannase [3]. Alves et al. [4] reported a novel magnetic composite from magnetite and azocasein to trypsin purification of fish Nile tilapia (crude extract). The obtainment of antioxidant peptides for use in food products was possible by immobilization of protease on magnetic support [5]. These are some of our research involving magnetic materials and their applications that we will address in more detail later.

1.1 Historical background: enzyme immobilization on magnetic particles

In 1916, the first scientific report on immobilization of enzymes was announced. This finding involved the invertase, a hydrolytic enzyme, which preserved their catalytic behavior after being absorbed on charcoal or aluminum hydroxide surface [6]. Robinson et al. [7] reported the first work of the use of magnetic particles as support to enzyme immobilization in 1973. The authors purposed two magnetic materials from iron oxide (magnetite) and another compound as a matrix to immobilize the enzymes: α-chymotrypsin and β-galactosidase for applications in bioreactors. A year later, Van Leemputten and Horisberger [8] immobilized trypsin and invertase on functionalized magnetite. Since then, magnetic separation has become an increasingly popular tool for the process of separating biological molecules and cells.

According to IUPAC gold book [9], an immobilized enzyme is defined as "a soluble enzyme bound to an insoluble organic or inorganic matrix, or encapsulated within a membrane in order to increase its stability and make possible its repeated or continued use." An efficient and robust immobilized derivative must preserve good retention of the catalytic activity, possess greater thermal and operational stability, be reused without considerable loss of activity, allow the easy separation of products and enzyme, and be resistant to microbial attack. In addition, a derivative immobilized on magnetic particles has advantages such as (i) easy and fast separation of the reaction medium by application of an external magnetic field, (ii) enzyme which is not stressed since conventional methods of separation (e.g., centrifugation and filtration) can be avoided, and (iii) large loading of biomolecules onto small particles (nanoparticles (NPs)) as a consequence of their high surface area.

Since the last decade, several scientific works about immobilization of enzymes have been published. In the first month of 2019, over 100 articles with the keyword "enzyme immobilization" have been reported in the PubMed database.

1.2 Immobilization strategies: Choosing the better approach

The choice of the immobilization method is as important as the nature of biomolecule (e.g., biochemical properties) and the experimental conditions chosen to obtain an immobilized derivative with desired features. So special attention should be given to immobilization approaches since the applicability of the immobilized derivative depends on this. The characteristics of the support are also relevant; however, it will be discussed later.

Overall, the methods of immobilization are categorized as irreversible and reversible since those interactions between enzyme and support are from weak physical adsorption to strong covalent bonds. Irreversible immobilization is understood as the attachment of the biocatalyst to the support with retention of the biological activity. However, the detachment of the biocatalyst will lead to the loss of its activity. Covalent bond, entrapment or microencapsulation, and cross-linking belong to this category. Already the adsorption and the affinity methodologies are

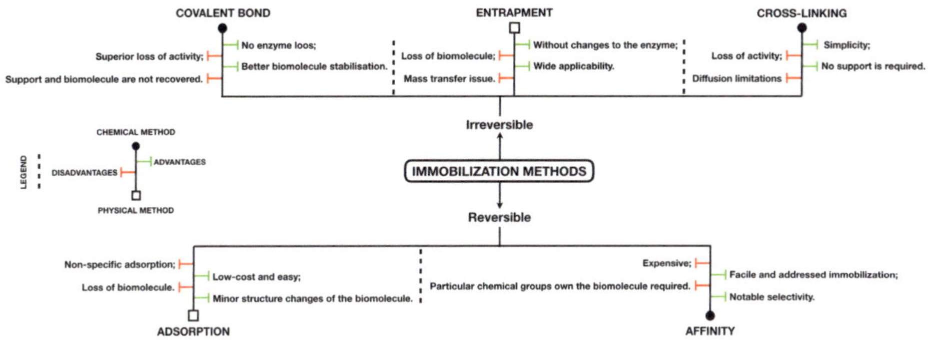

Figure 1.
Scheme with major strategies used for enzyme immobilization. Advantages and disadvantages are also highlighted.

associated with the reversible category. Additionally, the immobilization methods can be classified into chemical and physical according to support binding. **Figure 1** displays the most used immobilization methods together with some advantages and disadvantages.

Currently combinations of two or more immobilization methods can be employed to obtain derivative immobilized with features desired. Thus, the limitations from one type of immobilization could be avoided. Briefly, the main features of the immobilization methods will be mentioned as follows.

- **Covalent bond**. This method is based on the formation of covalent bonds between the biocatalyst and the support. For this, the presence of active chemical groups on surface of both components (i.e., biomolecule and support) is necessary. It is important to mention that these functional groups in the biocatalyst are not responsible for their catalytic activity. Moreover, the use of an activating agent, e.g., glutaraldehyde, which will lead the covalent attachment is required. Covalent bond is one of the most immobilization technique used.

- **Entrapment**. In this approach, the biocatalyst is restricted within a polymeric network. The substrate and products can reach the biocatalyst since their molecular weight is low. So, this method is suitable when mass transfer limitations through polymeric network are not a problem. Porous gels, fibers, and microencapsulation are some strategies to entrapping the biocatalyst. The immobilization by entrapment could be by essentially physical forces or include covalent bonding.

- **Cross-linking**. An arrangement of insoluble aggregates with high molecular weight is formed by a simple process involving the biocatalyst and bi- or multifunctional reagents or ligands. Due to this method which did not use a support and involve covalent bonds, conformation changes of the biocatalyst are possible leading to the loss of their activity. Glutaraldehyde, a bifunctional agent, is the most generally used for this immobilization technique. Cross-linked enzyme aggregates (CLEAs) have emerged as attractive alternative to produce physical aggregates with preservation of biocatalyst structure and hence their catalytic activity.

- **Adsorption**. This simple, fast, and inexpensive method leads to the formation of noncovalent interactions. Electrostatic adsorption is the linkage approach more used. Unfortunately, the interaction between the biocatalyst and the support may be modulated by some operational parameters such as ionic strength, pH, and temperature. Therefore, this approach is more appropriate when the physical adsorption of the biocatalyst is carried out in hydrophobic environments.

- **Affinity.** The immobilization by affinity interaction requires that the biocatalyst as well as the support present specific chemical groups on their surfaces, that is, the presence of complementary species, e.g., streptavidin-biotin interaction. As a consequence of the specific interaction, notable selectivity is a major benefit of this method. However, this procedure is expensive since antibodies or lectins can be used.

1.3 Why covalent immobilization?

There is not a universal method of immobilization and support for the enzymes and their uses in biotechnology. Several factors could drive the choice of a particular method, e.g., physicochemical properties of support or different features of substrates and products. As shown in **Figure 1**, all methods present advantages and disadvantages. However, why should the covalent bond be chosen? Covalent bonding is the most used strategy of the irreversible category. The removal of the covalent immobilized enzyme from the support without affecting its catalytic activity as an attempt to recover either the biomolecule or the support is meaningless due to hard process usually involved to disrupt the covalent bond. Moreover, better thermal and operational stability, as well as major resistance to pH, temperature, and solvent variations, are some of the well-known benefits of the immobilization approach. A good reason to choose this method is when a system with high stable protein coverage is recommended. Obtainment of the product of high purity, i.e., no contaminants including the enzyme, is another excuse to employ the covalent bond for immobilization.

2. About the particles used as support

In order to prepare an immobilized derivative, at least the biomolecule, the support, and the method of immobilization are required. Choosing the support is an essential step in the immobilization process since the characteristics of the material can influence in the performance of the biocatalyst. Even today there is no general rule for selecting the ideal support to attach the biomolecule. However, materials with characteristics such as chemical inertia, hydrophilic character, low cost, mechanical resistance, and resistance to microbial attack are widely used. Often in the literature, terms as matrix and carrier are found as synonyms of support.

Magnetic particles as a carrier to biomolecule immobilization are desirable materials due to easy separation of the biocatalyst from the reaction medium by application of an external magnetic field. Among the particles with magnetic property, the iron oxides, in particular, magnetite (Fe_3O_4), are the materials with notable uses in the immobilization process. Maghemite (γ-Fe_2O_3) and hematite (α-Fe_2O_3) are iron oxides derived from magnetite and can be found together with it.

2.1 Iron oxides: Why use it as a support of biomolecules?

Until now several magnetic particles obtained from pure iron oxide or a mixture of phases and composites materials have been presented as attractive support to enzyme immobilization or protein purification. To understand why these particles are so attractive, it might be useful to review the characteristics of the magnetite. The iron oxide is inorganic in nature, which presents a ratio of ferric/ferrous ions equal to 2:1, in stoichiometric magnetite. The ions positively charged (ferric, denoted as Fe^{3+}, and ferrous, expressed as Fe^{2+}) are distributed in two different sites with a crystal inverse spinel structure, formed by a cubic network face centered (fcc) of oxygen anions. Tetrahedral sites are filled by Fe^{3+} ions, while the octahedral sites are occupied by Fe^{3+} and Fe^{2+} ions [10].

According to the magnetic response, iron oxides can be classified as ferromagnetic, antiferromagnetic, paramagnetic, and ferrimagnetic. At room temperature, bulk magnetite is a ferrimagnetic material due to the combination of ferromagnetic and antiferromagnetic comportments that happens below the Curie temperature (850 K) [11]. This iron oxide presents a saturation magnetization (M_s) value between 92 and 100 emu g^{-1}. Maghemite is also a ferrimagnetic material (M_s = 60–80 emu g^{-1}), while the hematite is weakly ferromagnetic (M_s minor than 1 emu g^{-1}). It is important to mention that there is a correlation between the magnetic properties of the material with their size. So, when the magnetite presents nanometer size (below 15 nm), the superparamagnetic behavior is observed [12].

In addition to physicochemical and magnetic properties of the magnetite, using these particles as a carrier of the biomolecules could avoid laborious procedures such as decantation, centrifugation, or filtration.

2.2 Synthesis of iron oxides

At present, many synthesis methods to produce iron oxide particles have been developed. Evaluating the preparation method together with the features desires of the material (i.e., size, shape, size distribution, and surface chemistry) is very important once the magnetic property, as well as applications of material synthesized, depends on it. Moreover, the degree of structural defects (i.e., the presence of impurities) and distribution of the defects are also related to the synthesis methodology [13]. In general, the main preparation methods of the iron oxide particles can be classified as chemical, physical, and biological. Among these methodologies, the chemical route is the most used due to high yield along with low-cost production. Coprecipitation, hydrothermal, microemulsion, sonochemical, thermal decomposition, sol-gel synthesis, and electrochemical decomposition are some of the chemical methods [14]. A complete description about the most common preparation methods of the iron oxide particles together with their benefits and drawbacks can be found in the literature [1, 12, 14].

The "coprecipitation method" presents interesting characteristics that make it the chosen one among the chemical route category. For instance, this technique is simple and inexpensive and can be used for large-scale production. Briefly, iron oxide particles are produced by addition of a precipitating agent (e.g., ammonium hydroxide) to the solution containing a mixture of ferrous and ferric salts (e.g., $FeCl_2.4H_2O$ and $FeCl_3.6H_2O$) at room temperature or high temperature. A molar ratio equal to 2:1 for Fe^{3+}/Fe^{2+} ions is generally used. The obtainment of a black precipitate (Fe_3O_4) is a good indication that the synthesis was successful. The size, shape, composition and magnetic property of the particles can be modified depending on the iron precursors, iron concentrations, ferric/ferrous ratio, precipitating agent, pH, temperature, ionic strength, stirring velocity, surfactants addition, and working under inert atmosphere [10, 15].

Processes of the nucleation and the growth of crystals are involved in this technique. Both processes should take place separately, i.e., first the nucleation and then the growth. According to Wu et al. [10], pH is an important parameter to address toward the nucleation (solution pH minor than 11) and the growth (solution pH major than 11) of crystals. Even though the coprecipitation method is widely used to produce iron oxide particles with high saturation magnetization, some limitations like broad and large size distribution, poor crystallinity, aggregation, and tendency to oxidize can be cited. For instance, the formation of impurities (e.g., maghemite) depends on the initial and final solution pH as well as the reaction temperature [16]. Moreover, impurities are also observed in very small particles (minor than 20 nm) because of the high surface area/volume ratio, which allows a great number of surface atoms. This happening could lead to the formation of maghemite, for example, as a consequence of the oxidation of Fe^{2+} to Fe^{3+} ions. The maghemite can also be present when the magnetic sample is stored for an extensive period (6 months) as well as exposure to high temperatures (superior to 180°C) [16]. Since maghemite is also a ferrimagnetic material but with a slightly minor saturation magnetization than magnetite, the presence of maghemite in the magnetic sample may not be considered a disadvantage. However, the presence of hematite is an unwanted impurity since this oxide is weakly ferromagnetic and exhibits low saturation magnetization.

Our research group has used the coprecipitation method for the obtainment of magnetic particles to be used as a matrix for biomolecules immobilization [17, 18]. Among these materials, some magnetic composites were also synthesized. For this, materials (without magnetic property) can be added to the $FeCl_3.6H_2O/FeCl_2.4H_2O$ mixture before the magnetization aiming different purposes: magnetic particle composite synthesis.

Maciel et al. [19] synthesized a magnetic levan; that is, under a polymer of fructose, the iron oxide particles were formed. The magnetic levan particles were treated with sodium periodate ($NaIO_4$) and employed as a support for trypsin covalent immobilization.

Rêgo et al. [20] have reported a gum magnetic from *Parkia pendula* seeds as a matrix for concanavalin A (Con A) covalent immobilization. For this to be possible, seed gum was included in the solution containing the Fe^{3+} and Fe^{2+} ions. Afterward, it was functionalized with $NaIO_4$ allowing the covalent immobilization of the lectin Con A.

Mercês et al. [21] described the process to convert Dacron to magnetic Dacron-hydrazide (mDAC). Heparin (HEP) was activated by carbodiimide and N-hydroxysuccinimide and covalently linked to mDAC (mDAC-HEP). Human antithrombin was then purified by affinity chromatography using the mDAC-HEP.

Alves et al. [4] prepared a magnetic composite from azocasein and iron oxide particles (mAzo). The presence of azocasein, azo-dye-insoluble casein derivative, on the surface of the magnetic particles allowed trypsin to be purified by affinity binding. The enzyme forms complex with the modified substrate but does not hydrolyze it. Washing the enzyme-azocasein-magnetic particles removes unspecific proteins of the mixture. Afterward, the complex is disrupted by increasing the ionic strength, and the enzyme is collected in the supernatant.

Due to the inorganic nature of the iron oxides, the magnetic particles do not have chemical groups to enable biomolecules to bind covalently. Therefore, additional procedures should be performed for this purpose (functionalization). Two approaches can be carried out: (1) coating them with polymers containing these chemical groups and (2) adding materials encompassing these chemical arms during or after the magnetic particle synthesis (composites). In this context, Cabrera et al. [22] used the 3-aminopropyltriethoxysilane (APTES) as a silane agent to available amine groups on the surface of the magnetic diatomaceous earth (mDE). The treatment with APTES was carried out after the synthesis of the mDE particles. The composite material (mDE-APTES) showed efficiency as a matrix to immobilize invertase. Furthermore, Cabrera et al. [2] have also reported a simple, effective, and inexpensive synthesis methodology to obtain a magnetic composite made from mDE particles coated with polyaniline (mDE@PANI). The coating with PANI was carried out after preparation of the mDE particles. Three industrial enzymes (invertase, β-galactosidase, and trypsin) were successfully immobilized using glutaraldehyde as

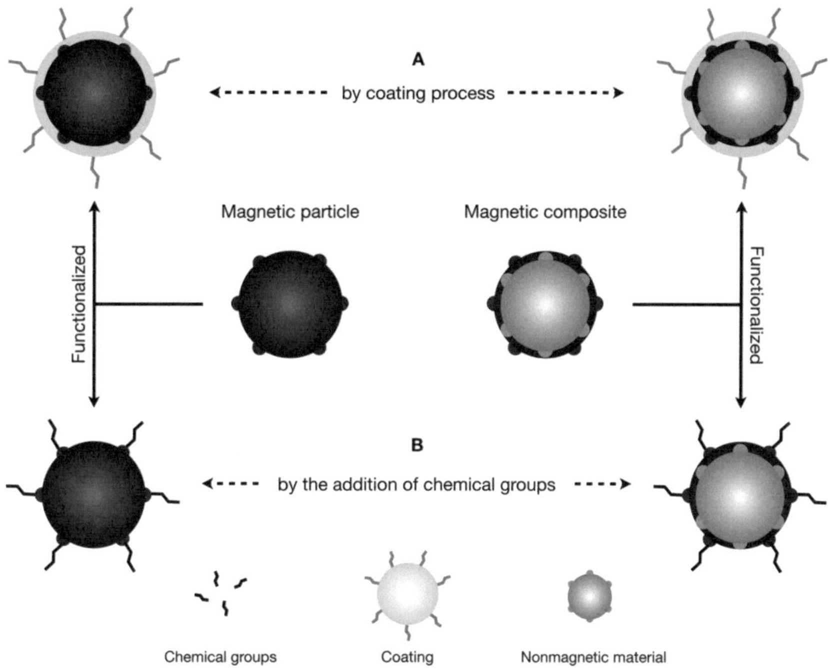

Figure 2.
Approaches to prepare functionalized magnetic particles. (A) Functionalized by coating process and (B) functionalized by the addition of chemical groups.

a chemical arm on the mDE@PANI particles. All magnetic bio-derivatives displayed superior performance (related to catalytic activity and stability) compared to the free enzyme. Thus, the mDE@PANI particles showed potential as a matrix to immobilize other biomolecules. Tannase was also covalently immobilized onto the mDE@PANI particles [3]. This enzyme catalyzes the hydrolysis of tannins present in several beverages. **Figure 2** displays the preparation of functionalized magnetic particles for biomolecule immobilization or protein purification.

2.3 Micro or nano iron oxide particles?

Before proceeding with the immobilization process, essential aspects such as material and surface science, biomolecule chemistry, and reaction engineering should be evaluated. A satisfactory immobilization will be achieved when all these aspects are properly integrated. However, this is not very easy due to the multidisciplinary nature of problems along with the main effects occurring in different length scales from nanometer to millimeter [23].

It is well known that an immobilized derivative presents, in the majority of times, minor retention of their catalytic activity when compared to its free counterpart. This happens for two reasons: (i) conformational changes of the biomolecule due to covalent bond with the support and (ii) problems arising of the catalytic reaction occur in a heterogeneous environment [24]. So, analyzing the behavior of biomolecule immobilized could help in choosing the most suitable particle size for the material used as support.

In the last times, the "nanoimmobilization" is widely used for several researchers to indicate nanostructures as support for the immobilization of biomolecules. The employment of NPs presents various benefits due to unique physical properties resulting from their nanometer size (below 100 nm). Higher surface area, significant biomolecule loading, superior mass transfer resistance, and minor diffusion problems are some of the advantages of the use of nanoparticles as a matrix. Also better stability and performance of biocatalyst immobilized along with a low protein unfolding were also reported [25, 26]. Some disadvantages for the NPs (e.g., large-scale application and price of material preparation) have been presented [26]. On the other hand, magnetic nanoparticles (MNPs) have been used as a support of several biological molecules as well as present superparamagnetism (i.e., magnetic response is only observed after application of magnetic field) and can be recovered by the use of a magnet [27]. So, the magnetic property as a plus feature to the nanoparticles would make them more attractive not only for potential uses in biotechnology [14] but also for a lot of biomedical applications including magnetic hyperthermia [28], drug delivery [29], contrast agent in magnetic resonance imaging (MRI) [30], and cellular therapy [31].

As above mentioned (Section 2.2), some factors influence the size (micro or nano) of iron oxide particles. For instance, Maciel et al. [17] assessed the effect of the temperature and the nature of the precipitating agent (strong base) to produce iron oxide nanoparticles. The authors employed sodium hydroxide (NaOH) as the precipitating agent and carried out the synthesis at low temperature (50°C). Small magnetic nanoparticles with a diameter near to 15 nm were obtained and used as a matrix to immobilize trypsin. The magnetic bio-derivative displayed about 90% retention of specific activity after five reuses. Despite Cabrera et al. [2] have reported mDE@PANI nanoparticles (~12 nm) as a promising matrix to immobilize trypsin, the immobilized derivative (mDE@PANI-TRYP) retained 75 and 60% of its initial activity after five and nine cycles of reusability, respectively. The decrease of catalytic activity could be attributed to the loss of the magnetic bio-derivative (mDE@PANI-TRYP) during the washing process after each reusability cycle. After coating with polyaniline, the mDE@PANI showed better stability in suspension. Thus, working with very small particles and good colloidal stability can lead to loss of the immobilized derivative.

Therefore, the choice of micro or nanoparticles as support will depend on several factors including colloidal stability of the particles, operating conditions, and application of the immobilized derivative, among others.

3. The chemistry of materials

In this section, a brief presentation of differential physicochemical techniques used to characterize the support and/or to demonstrate the efficiency of the immobilization methodology will be described. The description of theoretical basis, equipment, and conditions of analysis of the techniques are not the purpose of this material. The main information obtained from each analysis method as well as some examples will be presented below.

3.1 Magnetization measurements

In general, a material with magnetic property is analyzed by the magnetization measurement technique in order to quantify this property. The saturation magnetization (M_s), remanent magnetization (M_r), and coercivity (H_c) are among the main data obtained by this technique. Additionally, the presence of a hysteresis loop, as well as the M_r and H_c parameters, could help to assess the magnetic behavior of a material. For instance, a superparamagnetic material presents M_r and H_c values near to zero.

The inclusion of materials into the magnetic particles decreases their magnetization power although they still can be attracted by an external magnet. The saturation magnetization for the magnetic composite prepared from levan polymer and iron oxide particles was reduced tenfold. The happening can be attributed to the difficult alignment of magnetic dominions in the composite material due to the coating process. Furthermore, the addition of the levan increased the particles sizes varying from 20 to 60 μm for magnetite only and 100–200 μm for magnetic levan composite. The authors used this magnetic composite to immobilize trypsin by covalent binding [19].

Gregorio-Jauregui et al. [32] also showed a decrease in the saturation magnetization values as the amount of polymer (chitosan) in the particles was increased. For instance, bare magnetic nanoparticles (without chitosan—0 w/v%) and magnetic nanoparticles coated with chitosan (0.5 w/v%) presented a M_s near to 70 and 45 emu g^{-1}, respectively. Furthermore, the authors suggested that these findings could be associated with the direct relation between crystallinity and magnetization in magnetic particles. That is, magnetic materials with a good degree of crystallinity will present a large saturation magnetization. However, the addition of chitosan polymer (poor crystallinity) leads to a decrease of magnetic response.

Surface modification processes, including immobilization of enzyme, were evaluated by Defaei et al. [33]. The authors synthesized magnetic nanoparticles coated with silica and functionalized with naringin (MNP@SiO_2/NA). This material was employed as support to immobilize α-amylase (MNP@SiO_2/NA/AA). After each modification process a decrease on saturation magnetization values was observed due to the increase of thickness of the shell layer on the magnetic nanoparticles. So, the saturation magnetization values were 38, 27, and 22 emu g^{-1} for MNP@SiO_2, MNP@SiO_2/NA, and MNP@SiO_2/NA/AA, respectively.

3.2 X-ray diffraction (XRD)

X-ray diffraction (XRD) is an important method used for analyzing the intermolecular structure of ordered materials. However, this technique is not appropriate for quantifying the degree of order. Magnetic materials such as bare iron oxide particles as well as magnetic composites can be characterized by XRD analysis in order to evaluate the presence of different components in the sample. XRD can also be used to estimate the particle size by using the Scherrer equation, for example.

By using XRD technique, to differentiate between magnetite and maghemite is not possible since the iron oxides present similar standard XRD patterns. According to the International Center of Diffraction Data (reference code: ICDD 019-0629), the crystal planes at (111), (220), (311), (400), (422), (511), (440), (620), and (533) corresponding to the 2θ peaks at 18.44, 30.30, 35.67, 43.37, 53.80, 57.35, 62.97, 71.43, and 74.48° are attributed to both magnetite and maghemite [17]. For instance, Gregorio-Jauregui et al. [32] could not differentiate by XRD technique the iron oxides present in magnetic nanoparticles coated with chitosan. The authors attributed the presence of magnetite due to the black color of the magnetic composite. Furthermore, the coating with chitosan did not affect the crystalline structure of the magnetic nanoparticles.

Cabrera et al. [2] assessed by XRD the chemical composition (qualitative data) as well as the crystalline structure of magnetic diatomaceous earth coated with polyaniline (mDE@PANI) nanoparticles. The XRD pattern of the mDE@PANI sample displayed characteristic peaks for crystalline and amorphous silica along with albite, polyaniline, and magnetite. The iron oxide was the predominant crystalline phase. Additionally, the authors reported that the coating process with PANI did not affect the crystallinity degree of the magnetic sample since the narrow peaks were preserved.

Díaz-Hernández et al. [34] reported the use of magnetite nanoparticles coated with chitosan (Fe_3O_4@chitosan) as support for immobilization of enzymes. In spite of XRD pattern which displayed a low peak at 18° probably related to maghemite, the authors concluded that magnetite was present in the Fe_3O_4@chitosan nanoparticles. Therefore, the XRD technique revealed that the addition of chitosan polymer did not affect the crystal structure of the magnetic sample. Moreover, the authors carried out the XRD spectrum for enzyme immobilized by cross-linking. The XRD pattern for the immobilized derivative displayed broad peaks and with low intensity, but all peaks were in agreement with magnetite. This finding could be attributed to the amorphous nature of the biomolecule immobilized.

3.3 Mössbauer spectroscopy (MS)

Mössbauer spectroscopy (MS) is a sensitive technique to the iron ionic state and environments. MS can distinguish the ferric (Fe^{3+}) and ferrous (Fe^{2+}) ions because of their different isomer shifts. Thus, magnetite, maghemite, and hematite, for example, can be detected in a magnetic sample. Since magnetite mainly presents potential biotechnological and biomedical applications, it is very important to know the main iron oxide phases present in the sample. Therefore, the MS technique is very useful for monitoring the processes of preparation and modification of the iron oxides at different sizes. However, small magnetic nanoparticles (below than 10 nm) must be analyzed at low temperatures in order to block the superparamagnetic relaxation of them [35].

Cabrera et al. [36] proposed two magnetic composites as a matrix to immobilize invertase via covalent bonding. For this, clay minerals such as montmorillonite (MMT) and diatomaceous earth (DE) were used as a nonmagnetic component. Using MS technique, it was possible to evaluate the main iron oxide phases present in the magnetic composites (mMMT and mDE). Mössbauer results revealed a mixture of magnetite and maghemite in equal proportion for the mMMT particles, while a pure magnetite phase was observed in the mDE particles.

Storage conditions can also lead to changes in magnetic property due to phase transformation of iron oxides. Rümenapp et al. [37] described strong oxidation of the bare magnetic nanoparticles on the fourth day of preparation. The complete oxidation to maghemite was observed in the fourth week. In order to avoid the oxidation process, the authors suggested a coating process on the magnetic surface. Iron oxide nanoparticles without and with polyaniline (PANI) coating were assessed by Maciel et al. [17] using MS analysis. The authors described the presence of maghemite in the two samples. In addition, the coating with PANI did not change the chemical nature of the magnetic sample. Similarly, Cabrera et al. [2] functionalized with PANI a magnetic composite from diatomaceous earth and proposed it as promising support to enzyme immobilization. Due to the great catalytic performance of the magnetic bio-derivatives, the authors evaluated the magnetic behavior by MS technique. It is not common to analyze the magnetic sample containing the biomolecule may be due to the small amount of biomolecule immobilized in most of the time. MS results showed the magnetite as the major iron oxide phase in the magnetic composites (mDE and mDE@PANI). Moreover, the hematite was not detected in the samples.

3.4 Scanning electronic microscopy (SEM) and transmission electronic microscopy (TEM)

The structural and morphology characterization can be performed by scanning electronic microscopy (SEM) and transmission electronic microscopy (TEM). SEM can be employed to investigate the morphological surface of the support as well as the biomolecule. However, information related to the internal structure of the sample cannot be obtained by this technique. Additionally, SEM also provides the particle size. TEM is mainly used to determine the particle size, size distribution, and particle shape and to evaluate the effectiveness of the functionalization process (e.g., coating with a polymer) along with the thickness of the coating. In general, the particle size values found by TEM are in agreement with those obtained by XRD.

Cabrera et al. [2] used SEM technique to display physical changes in the size and surface of diatomaceous earth (DE) after the magnetization and functionalization processes. Strong operating conditions such as high temperature could lead to the destruction of DE frustules along with a rougher surface. A surface modification as the coating process with PANI on a magnetic core along with the morphology, shape, and size of material can be evaluated by using TEM. Maciel et al. [17] described a synthesis methodology to obtain magnetic nanoparticles coated with PANI as a matrix to immobilize trypsin. TEM results displayed a magnetic system with a spherical shape and particle size near to 15 nm.

Defaei et al. [33] demonstrated by SEM and TEM techniques the immobilization of α-amylase on magnetic nanoparticles. A slight increase in the particle size after the immobilization process suggested the effective enzyme immobilization. Moreover, the presence of oxygen, sulfur, and nitrogen atoms in the sample containing the enzyme by energy-dispersive X-ray spectroscopy (EDX) confirmed the presence of the biomolecule.

Gregorio-Jauregui et al. [32] reported the use of scanning transmission electron microscopy (STEM) technique to determine the size and morphology of magnetic nanoparticles coated with chitosan. The magnetic system presented a small size (9.9–11 nm) and spherical morphology. The STEM analysis combines the principles used by both SEM and TEM techniques, and it can also be used to locate biomolecules as well as active groups inside the matrix. For this, a high-angle annular dark field (HAADF) detector is coupled to STEM since the contrast is associated with the atomic number [38]. Mayoral et al. [39] demonstrated the presence of lipase immobilized inside the pores of silica by STEM-HAADF.

3.5 Thermal gravimetric analysis (TGA)

Thermal gravimetric analysis (TGA) is a thermal analysis technique which provides quantitative data (i.e., thermal stability, reaction rates, oxidation process, components quantification, and kinetic of decomposition). In general, TGA is used to study changes in physical and chemical properties, e.g., oxidation, dehydration, and decomposition, as a function of temperature [40]. Thermal stability of magnetic particles used as a matrix for immobilization of biomolecules is evaluated by weight loss (decomposition process) of the sample. An additional process (functionalization) carried out on the surface of magnetic particles can also be observed by this technique. That is, the amount of organic material in a sample can be detected and quantified. For instance, Aydemir and Güler [41] have reported the use of magnetic bio-derivative (laccase immobilized on magnetic chitosan-clay composite) for phenol removal. Structural characterization of the magnetic composite by TGA revealed superior thermal stability of chitosan after addition of clay and magnetite nanoparticles. Moreover, at 230°C the composite (magnetic chitosan-clay beads) presented a weight loss of 16% corresponding to the decomposition and elimination of the polymeric component. Neri et al. [42] used magnetic polysiloxane-polyaniline (mPOS-PANI) particles as support to immobilize β-galactosidase. TGA technique was employed to evaluate the coating process with PANI and thermal resistance of the mPOS (without PANI) and mPOS-PANI particles. Thermal degradation of PANI was observed near to 325°C. The magnetic samples (mPOS and mPOS-PANI) displayed a remaining weight loss probably due to the removal of the solvent.

Technique	Information type	Parameter[a]	Easy access to technique?
Magnetization measurement	Quantitative	Saturation magnetization, remanent magnetization, and coercivity	Yes
X-ray diffraction (XRD)	Qualitative and quantitative	Chemical composition, crystallinity degree, crystalline phase, and crystalline size	Yes
Mössbauer spectroscopy (MS)	Qualitative	Specific to differentiate between iron oxide phases (e.g., magnetite, maghemite, and hematite)	No
Electronic microscopy (SEM and TEM)	Qualitative and quantitative	Particle size, morphology, size distribution, and shape	Yes
Thermal gravimetric analysis (TGA)	Quantitative	Mass change in processes such as oxidation, decomposition, and dehydration	Yes

[a]*Some parameters provided by the technique.*

Table 1.
Main features of some physicochemical techniques to characterize a magnetic material or magnetic bio-derivative.

TGA can also be employed to determine the degree of functionalization of the support, the effectiveness of the immobilization method, and structural information of biomolecule after the immobilization [40]. An attractive nanobiocatalyst consisting of α-amylase (AA) immobilized on magnetic nanoparticles (MNP@SiO_2) functionalized with naringin (NA) was proposed [33]. The functionalization and immobilization processes were assessed by TGA analysis. The results for the MNP@SiO_2/NA (without enzyme) and MNP@SiO_2/NA/AA (with enzyme) showed that the major weight loss was associated with the removal of the organic moieties. Moreover, the difference in weight loss between these samples was used to evidence the α-amylase immobilization.

In order to conclude this section, **Table 1** exhibits a summary with the main information about the physicochemical techniques above described. The reader can use the information contained therein to evaluate the desired parameters. It is important to mention that these techniques can be applied in both magnetic materials with and without the biomolecule. Moreover, other techniques can also be performed to characterize a magnetic bio-derivative.

4. Biotechnological applications

The immobilization of different biomolecules in magnetic material has been used for several biotechnological applications in biomedicine [5], environment [43], and food industry [22] (**Figure 3**). At present, magnetic bio-derivatives have found potential uses in biodiesel production. Knowing it as a renewable, biodegradable, noninflammable, and nontoxic product, the biodiesel has been studied as an attractive substitute to the diesel fuel based on petroleum. Lipase immobilized on magnetic particles has been used to produce biodiesel by enzymatic transesterification [44, 45]. Moreover, magnetic microreactors employing immobilized enzymes are widely investigated to study and manipulate bioprocesses. For instance, the prediction of pharmaceutical response in animal models could be avoided by using enzymatic microreactors in culture systems. The authors advice reading the review article to find more information about magnetic enzyme microreactors [46].

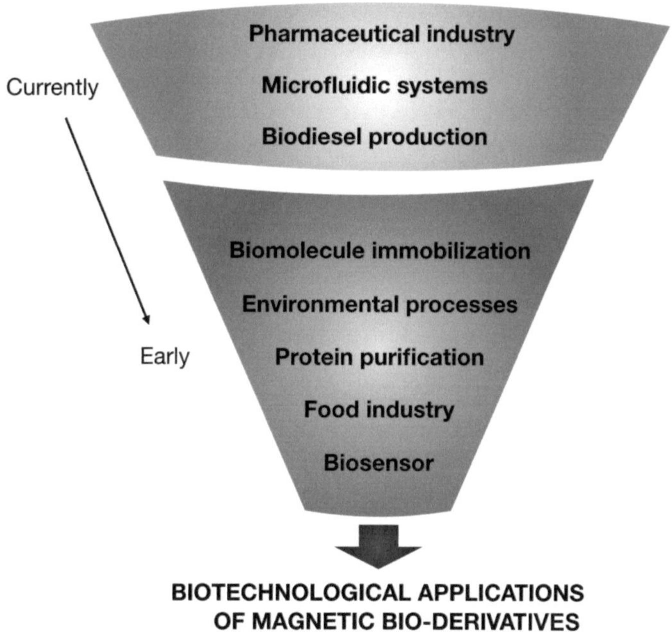

Figure 3.
Major biotechnological uses of the magnetic bio-derivatives.

Biocompatibility, specificity, ability to recognize other molecules, and operational stability presented by biomolecules such as enzymes [2], carbohydrates [21], antibodies [47], antigens [48], peptides [49], DNA [50], and glycoprotein [20], among others, are properties that make biomolecules attractive to biotechnological applications. The possibility of reuse, the easy removal of the magnetic bio-derivative from the reaction medium, the greater stability of the biomolecule, the high productivity, and the low-cost of production are some of the advantages of the immobilization.

Considering the above, the topics below exhibit the most recent developments in our research group. Applications such as trypsin purification, obtaining of antioxidant peptides, affinity purification of antithrombin, high sucrolytic activity by invertase, and purification of glycoproteins are discussed.

4.1 Trypsin purification using magnetic particles of azocasein-iron composite

The purpose of this work was to develop a trypsin purification strategy based on affinity binding with ferromagnetic particles of azocasein composite (mAzo) [4]. Trypsin was purified when the fish crude extract (2 mL) obtained from intestines of fish Nile tilapia (*Oreochromis niloticus*) was exposed to the magnetic bio-derivate (mAzo: 100 mg) for 2 h, removed from the reaction medium with the aid of a magnetic field, and washed with buffer (0.1 M Tris-HCl) seven times to remove unbound protein and three times under high ionic strength (3 M NaCl) to leach off the protein. The specific activity of the free enzyme present in the crude extract was 60-fold lower than that of the preparation. The optimum performance suggests that the mAzo composite, besides being reused, can be applied to purify trypsin from other sources.

4.2 Optimization of *Penicillium aurantiogriseum* protease immobilization on magnetic nanoparticles for antioxidant peptides' obtainment

In this work, magnetic nanoparticles coated with polyaniline were used to immobilize protease from *Penicillium aurantiogriseum* and applied to the production of antioxidant peptides derived from bovine casein [5]. The casein was hydrolyzed using the magnetic bio-derivative, uncovering its peptides that were sequenced and had antioxidant properties tested through 2,2′-azino-bis (3-ethylbenzothiazoline-6-sulfonic acid) diammonium salt (ABTS) radical scavenging and hydrogen peroxide scavenging assays. After being reused for five times, the magnetic bio-derivative was able to maintain more than 74% of the initial activity. Hydrolyzed casein presented similar peptide when used; the free and immobilized enzyme and prevalent peptides could be sequenced. When comparing the reactive oxygen species (ROS) scavenging activity, the hydrolysates presented 2.5 times more than nonhydrolyzed casein, which allows the use of an immobilized protease to obtain natural ingredients derived from casein attractive, indicating potential use in the production of functional foods.

4.3 Synthesis and characterization of magnetized Dacron-heparin composite employed for antithrombin affinity purification

The focus of this work was to produce a Dacron magnetic composite with heparin (mDAC-HEP) for the use in the purification of antithrombin from human plasma [21]. To obtain mDAC-HEP, heparin was activated by carbodiimide and N-hydroxysuccinimide, allowing covalent attachment. Purified antithrombin was released from the magnetic bio-derivative using solutions with increasing ionic strength (NaCl). The affinity properties of mDAC-HEP after 2 years of storage were preserved, and the magnetic bio-derivative was able to be reused for at least tenfold. The presence of the expected antithrombin size (58 kDa) was revealed in the bands by electrophoresis of the eluates. The composite synthesis was considered easy, low-cost, magnet-based affinity purification steps, and reusable.

4.4 High sucrolytic activity by invertase immobilized onto magnetic diatomaceous earth nanoparticles

The purpose of this work was to produce magnetic diatomaceous earth nanoparticles for invertase immobilization (mDE-APTES-invertase) using an easy and low-cost method that could offer high sucrolytic activity [22]. To obtain the results of high residual specific activity (92.5%), an experimental design was made with the objective of achieving the best immobilization conditions. This activity was able to be 2.42 times higher than other derivatives reported in the literature. The thermal and storage stability of the immobilized invertase was verified, and after 120 days of storage, the enzymatic derivative retained 80% of the activity, whereas the free enzyme lost practically all the activity. After ten reuses mDE-APTES-invertase retained 60% of residual activity. Considering the ease of obtaining the matrix and its efficiency, this nanocomposite proves promising to immobilize invertase from different origins and other biomolecules.

4.5 Magnetic *Parkia pendula* seed gum as a matrix for concanavalin A lectin immobilization and its application in affinity purification

The main objective of this work was to obtain a magnetic matrix with *Parkia pendula* seed gum to covalently immobilize concanavalin A [20]. The application of this magnetic composite was to obtain glycoconjugates through affinity purification. The obtained gum in the process was magnetized and activated with $NaIO_4$. Concanavalin A immobilized on the magnetic composite was used for the recognition of bovine serum fetuin glycoprotein. A glucose solution (300 mM) was used to carry out the election of the fetuin, and confirmation was made via SDS-PAGE.

A lectin immobilization efficiency of 63% was achieved and 14% fetuin purification. The magnetic composite is promising, given the results obtained, to be used with magnetic polysaccharide matrix to immobilize other lectins. Because it is a magnetic system, it can be used for affinity purification, and its recovery is performed easily and quickly with the help of a magnetic field.

5. Conclusions

This chapter has gathered primary and interesting information about immobilization of biomolecules, how to choose the best immobilization methodology, the support for the immobilization, and, among the magnetic materials, why iron oxide particles are highly used as a matrix to immobilization. Several physicochemical techniques were mentioned and described highlighting their features and the main information provided. Lastly, we presented some magnetic bio-derivatives with potential biotechnological applications which could be applied in other areas. So the authors hope that this material has been useful not only for the enzymologists but also to scientists working with applied surface science.

Acknowledgements

This work was financially supported by the Brazilian Agencies: FACEPE (Fundação de Amparo a Ciência e a Tecnologia do Estado de Pernambuco; grant number BFP-0132-1.05/17), CNPq (Conselho Nacional de Desenvolvimento Científico e Tecnológico; grant number 301830/2017-7), and CAPES (Coordenação de Pessoal de Nível Superior; grant number 23038.007659/2011-26).

Conflict of interest

The authors confirm that there are no conflicts of interest in this work.

Author details

Mariana Paola Cabrera[1*], David Fernando Morais Neri[2], Fernando Soria[3] and Luiz Bezerra Carvalho Jr [1]

1 Departamento de Bioquímica and Laboratório de Imunopatologia Keizo Asami, Universidade Federal de Pernambuco, Cidade Universitária, Recife, PE, Brazil

2 Universidade Federal do Vale do São Francisco, Petrolina, PE, Brazil

3 Departamento de Química, Universidad Nacional de Salta, Salta, Argentina

*Address all correspondence to: marian.br09@gmail.com

References

[1] Arias SL, Pessan PJ, Vieira PA, Lima MT, Delbem CA, Monteiro RD. Iron oxide nanoparticles for biomedical applications: A perspective on synthesis, drugs, antimicrobial activity, and toxicity. Antibiotics. 2018;**7**(2):46

[2] Cabrera MP, TFd F, RVBd S, CRDd A, Quispe Marcatoma J, Maciel JC, et al. Polyaniline-coated magnetic diatomite nanoparticles as a matrix for immobilizing enzymes. Applied Surface Science. 2018;**457**:21-29

[3] de Lima JS, Cabrera MP, de Souza Motta CM, Converti A, Carvalho LB. Hydrolysis of tannins by tannase immobilized onto magnetic diatomaceous earth nanoparticles coated with polyaniline. Food Research International. 2018;**107**:470-476

[4] Alves MHME, Nascimento GA, Cabrera MP, Silvério SIC, Nobre C, Teixeira JA, et al. Trypsin purification using magnetic particles of azocasein-iron composite. Food Chemistry. 2017;**226**:75-78

[5] Duarte Neto JMW, Maciel JC, Campos JF, LBd CJ, Marques DAV, Lima CA, et al. Optimization of *Penicillium aurantiogriseum* protease immobilization on magnetic nanoparticles for antioxidant peptides' obtainment. Preparative Biochemistry and Biotechnology. 2017;**47**(7):644-654

[6] Homaei AA, Sariri R, Vianello F, Stevanato R. Enzyme immobilization: An update. Journal of Chemical Biology. 2013;**6**(4):185-205

[7] Robinson PJ, Dunnill P, Lilly MD. The properties of magnetic supports in relation to immobilized enzyme reactors. Biotechnology and Bioengineering. 1973;**15**(3):603-606

[8] Van Leemputten E, Horisberger M. Immobilization of enzymes on magnetic particles. Biotechnology and Bioengineering. 1974;**16**(3):385-396

[9] McNaught AD, Wilkinson AIUPAC. Compendium of Chemical Terminology, 2nd ed. (the "Gold Book"). Oxford: Blackwell Scientific Publications; 1997

[10] Wu W, Wu Z, Yu T, Jiang C, Kim W-S. Recent progress on magnetic iron oxide nanoparticles: synthesis, surface functional strategies and biomedical applications. Science and Technology of Advanced Materials. 2015;**16**(2):023501

[11] Blaney L. Magnetite (Fe_3O_4): Properties, Synthesis, and Applications. In: The Lehigh Review. Bethlehem: Lehigh Preserve; 2007;**15**

[12] Assa F, Jafarizadeh-Malmiri H, Ajamein H, Anarjan N, Vaghari H, Sayyar Z, et al. A biotechnological perspective on the application of iron oxide nanoparticles. Nano Research. 2016;**9**(8):2203-2225

[13] Akbarzadeh A, Samiei M, Davaran S. Magnetic nanoparticles: preparation, physical properties, and applications in biomedicine. Nanoscale Research Letters. 2012;**7**(1):144

[14] Ali A, Zafar H, Zia M, Ul Haq I, Phull AR, Ali JS, et al. Synthesis, characterization, applications, and challenges of iron oxide nanoparticles. Nanotechnology, Science and Applications. 2016;**9**:49-67

[15] Laurent S, Forge D, Port M, Roch A, Robic C, Vander Elst L, et al. Magnetic iron oxide nanoparticles: Synthesis, stabilization, vectorization, physicochemical characterizations, and biological applications. Chemical Reviews. 2008;**108**(6):2064-2110

[16] Gnanaprakash G, Mahadevan S, Jayakumar T, Kalyanasundaram P,

Philip J, Raj B. Effect of initial pH and temperature of iron salt solutions on formation of magnetite nanoparticles. Materials Chemistry and Physics. 2007;**103**(1):168-175

[17] Maciel JC, Mercês D, AA CM, Shigeyosi WT, de Souza SD, Olzon-Dionysio M, et al. Magnetic nanoparticles coated with polyaniline to stabilize immobilized trypsin. Hyperfine Interactions. 2016;**237**(1):3

[18] Neri DFM, Balcão VM, Dourado FOQ, Oliveira JMB, Carvalho LB, Teixeira JA. Immobilized β-galactosidase onto magnetic particles coated with polyaniline: Support characterization and galactooligosaccharides production. Journal of Molecular Catalysis B: Enzymatic. 2011;**70**(1):74-80

[19] Maciel JC, Andrad PL, Neri DFM, Carvalho LB, Cardoso CA, Calazans GMT, et al. Preparation and characterization of magnetic levan particles as matrix for trypsin immobilization. Journal of Magnetism and Magnetic Materials. 2012;**324**(7):1312-1316

[20] Rêgo MJBM, Almeida SM, Bezerra SA, Carvalho Júnior LB, Beltrão EIC. Magnetic Parkia pendula seed gum as matrix for Concanavalin A lectin immobilization and its application in affinity purification. Annals of the Brazilian Academy of Sciences. 2014;**86**(3):1351-1357

[21] AADd M, Silva RS, Silva KJS, Maciel JC, Oliveira GB, Buitrago DM, et al. Synthesis and characterisation of magnetised Dacron-heparin composite employed for antithrombin affinity purification. Journal of Chromatography B. 2016;**1038**:73-79

[22] Cabrera MP, Assis CRD, Neri DFM, Pereira CF, Soria F, Carvalho LB. High sucrolytic activity by invertase immobilized onto magnetic diatomaceous earth nanoparticles. Biotechnology Reports. 2017;**14**:38-46

[23] Bolivar JM, Eisl I, Nidetzky B. Advanced characterization of immobilized enzymes as heterogeneous biocatalysts. Catalysis Today. 2016;**259**:66-80

[24] Rodrigues RC, Ortiz C, Berenguer-Murcia Á, Torres R, Fernández-Lafuente R. Modifying enzyme activity and selectivity by immobilization. Chemical Society Reviews. 2013;**42**(15):6290-6307

[25] Gupta MN, Kaloti M, Kapoor M, Solanki K. Nanomaterials as matrices for enzyme immobilization. Artificial Cells, Blood Substitutes, and Biotechnology. 2011;**39**(2):98-109

[26] Cipolatti EP, Silva MJA, Klein M, Feddern V, Feltes MMC, Oliveira JV, et al. Current status and trends in enzymatic nanoimmobilization. Journal of Molecular Catalysis B: Enzymatic. 2014;**99**:56-67

[27] Leitgeb M, Knez Ž, Vasić K. Micro- and nanocarriers for immobilization of enzymes. In: Stanciu SG, editor. Micro and Nanotechnologies for Biotechnology. IntechOpen; 2016. DOI: 10.5772/63129

[28] Piñeiro Y, Vargas Z, Rivas J, López-Quintela MA. Iron oxide based nanoparticles for magnetic hyperthermia strategies in biological applications. European Journal of Inorganic Chemistry. 2015;**2015**(27):4495-4509

[29] Vangijzegem T, Stanicki D, Laurent S. Magnetic iron oxide nanoparticles for drug delivery: Applications and characteristics. Expert Opinion on Drug Delivery. 2019;**16**(1):69-78

[30] Wang Y, Xu C, Chang Y, Zhao L, Zhang K, Zhao Y, et al. Ultrasmall superparamagnetic iron oxide nanoparticle for T2-weighted magnetic resonance imaging. ACS Applied Materials & Interfaces. 2017;**9**(34):28959-28966

[31] Azevedo-Pereira RL, Rangel B, Tovar-Moll F, Gasparetto EL, Attias M, Zaverucha-do-Valle C, et al. Superparamagnetic iron oxide nanoparticles as a tool to track mouse neural stem cells in vivo. Molecular Biology Reports. 2018;**46**:191-198

[32] Gregorio-Jauregui KM, Pineda GM, Rivera-Salinas JE, Hurtado G, Saade H, Martinez JL, et al. One-step method for preparation of magnetic nanoparticles coated with chitosan. Journal of Nanomaterials. 2012;**2012**:8. Article ID: 813958

[33] Defaei M, Taheri-Kafrani A, Miroliaei M, Yaghmaei P. Improvement of stability and reusability of α-amylase immobilized on naringin functionalized magnetic nanoparticles: A robust nanobiocatalyst. International Journal of Biological Macromolecules. 2018;**113**:354-360

[34] Díaz-Hernández A, Gracida J, García-Almendárez BE, Regalado C, Nuñez R, Amaro-Reyes A. Characterization of magnetic nanoparticles coated with chitosan: A potential approach for enzyme immobilization. Journal of Nanomaterials. 2018;**2018**:11

[35] Dézsi I, Fetzer C, Gombkötő Á, Szűcs I, Gubicza J, Ungár T. Phase transition in nanomagnetite. Journal of Applied Physics. 2008;**103**(10):104312

[36] Cabrera M, Maciel JC, Quispe-Marcatoma J, Pandey B, Neri DFM, Soria F, et al. Magnetic composites from minerals: Study of the iron phases in clay and diatomite using Mössbauer spectroscopy, magnetic measurements and XRD. Hyperfine Interactions. 2014;**224**(1):197-204

[37] Rümenapp C, Wagner FE, Gleich B. Monitoring of the aging of magnetic nanoparticles using Mössbauer spectroscopy. Journal of Magnetism and Magnetic Materials. 2015;**380**:241-245

[38] Mayoral A, Carey T, Anderson PA, Lubk A, Diaz I. Atomic resolution analysis of silver ion-exchanged zeolite A. Angewandte Chemie. 2011;**123**(47):11426-11429

[39] Mayoral A, Arenal R, Gascón V, Márquez-Álvarez C, Blanco RM, Díaz I. Designing functionalized mesoporous materials for enzyme immobilization: Locating enzymes by using advanced TEM techniques. ChemCatChem. 2013;**5**(4):903-909

[40] Mohamad NR, Marzuki NHC, Buang NA, Huyop F, Wahab RA. An overview of technologies for immobilization of enzymes and surface analysis techniques for immobilized enzymes. Biotechnology & Biotechnological Equipment. 2015;**29**(2):205-220

[41] Aydemir T, Güler S. Characterization and immobilization of Trametes versicolor laccase on magnetic chitosan-clay composite beads for phenol removal. Artificial Cells, Nanomedicine, and Biotechnology. 2015;**43**(6):425-432

[42] Neri DFM, Balcão VM, Dourado FOQ, Oliveira JMB, Carvalho LB, Teixeira JA. Galactooligosaccharides production by β-galactosidase immobilized onto magnetic polysiloxane-polyaniline particles. Reactive and Functional Polymers. 2009;**69**(4):246-251

[43] Darwesh OM, Matter IA, Eida MF. Development of peroxidase enzyme immobilized magnetic nanoparticles for bioremediation of textile wastewater dye. Journal of Environmental Chemical Engineering. 2019;**7**(1):102805

[44] Miao C, Yang L, Wang Z, Luo W, Li H, Lv P, et al. Lipase immobilization on amino-silane modified superparamagnetic Fe_3O_4 nanoparticles as biocatalyst for biodiesel production. Fuel. 2018;**224**:774-782

[45] Karimi M. Immobilization of lipase onto mesoporous magnetic

nanoparticles for enzymatic synthesis of biodiesel. Biocatalysis and Agricultural Biotechnology. 2016;**8**:182-188

[46] Gkantzou E, Patila M, Stamatis H. Magnetic microreactors with immobilized enzymes—From assemblage to contemporary applications. Catalysts. 2018;**8**(7):282

[47] Foubert A, Beloglazova NV, Hedström M, De Saeger S. Antibody immobilization strategy for the development of a capacitive immunosensor detecting zearalenone. Talanta. 2019;**191**:202-208

[48] Puttharugsa C, Wangkam T, Houngkamhang N, Yodmongkol S, Gajanandana O, Himananto O, et al. A polymer surface for antibody detection by using surface plasmon resonance via immobilized antigen. Current Applied Physics. 2013;**13**(6):1008-1013

[49] González I, Oliver-Ortega H, Tarrés Q, Delgado-Aguilar M, Mutjé P, Andreu D. Immobilization of antimicrobial peptides onto cellulose nanopaper. International Journal of Biological Macromolecules. 2017;**105**:741-748

[50] Tang L, Guo Z, Zheng X. Increasing target hybridization kinetics by immobilizing DNA probes at Au nanoflower ultramicroelectrode. Sensors and Actuators B: Chemical. 2019;**279**:1-6

Chapter 3

Surface Science Engineering through Sol-Gel Process

Cristian Carrera-Figueiras, Yamile Pérez-Padilla, Manuel Alejandro Estrella-Gutiérrez, Erbin G. Uc-Cayetano, Juan Antonio Juárez-Moreno and Alejandro Avila-Ortega

Abstract

Sol-gel synthesis is used to obtain coatings that can modify the surfaces of metals to avoid corrosion or to enhance the biocompatibility and bioactivity of metals and their alloys that are of biomedical interest. Anticorrosion coatings composed of smart coatings and self-healing coatings will be described. TiO_2, hydroxyapatite, bioglass, and hybrid coatings synthetized by sol-gel technology will be briefly introduced with regard to their role in surface-modifying metals for biomedical purposes. Finally, although there are other approaches to surface-modifying metals for either anticorrosion or biomedical purposes, sol-gel methods have several advantages in controlling surface chemistry composition and functionality.

Keywords: surface modification, coatings, sol-gel, corrosion, biomaterials

1. Introduction

The need to control interactions between materials and their surrounding environment is mainly concerned with the surface properties of these materials. In this sense, the sol-gel process is being increasingly used to surface modify a wide range of materials such as metals, organic polymers, inorganic particles, and glasses, where wettability, biocompatibility, porosity, corrosion, catalytic activity, and selective adsorption of analytes can be controlled at the required substrate surfaces. For this reason, sol-gels have gained attention in different scientific and technological fields, including metallurgy, biomaterials, analytical chemistry, and photocatalysts. In this chapter, an overview of sol-gel chemistry and its applicability to modifying the surfaces of metals and their alloys for anticorrosion and biomedical purposes will be presented.

2. Sol-gel chemistry

The sol-gel technology is a technique that has been widely employed in the synthesis of inorganic polymers or advanced organic-inorganic hybrids due to versatile and simplicity. Two chemical reactions are involved, hydrolysis and condensation, and these produce a variety of organic-inorganic networks from precursor monomers such as silicon alkoxide or metals alkoxides [1, 2]. This technique permit to obtain materials in any form and to it is possible to produce homogeneous materials with the desirable properties of toughness, high purity, optical transparency, chemical stability, controlled porosity, and thermal resistance at room temperature and low cost [3].

The sol-gel process, as the name implies, is the transition of a liquid colloidal solution (sol) to a solid three-dimensional matrix (gel). The precursors for the synthesis of these colloids consist of a metal or metalloid element surrounded by several reactive ligands. Metal alkoxides are the most popular, since they react easily with water. The most extensively used are the metal alkoxides and the alkoxysilanes, such as tetramethoxysilane (TMOS) and tetraethoxysilane (TEOS). Nevertheless, other alkoxides such as aluminates, titanates, and borates are also commonly used in the sol-gel process, where they are often mixed with TEOS [1].

The most widely used method is the organic approach, which generally starts with a solution of monomeric metal or metalloid alkoxide precursors, $M(OR)_n$, in an alcohol or another low-molecular-weight organic solvent. Here, M represents a network-forming element such as Si, Ti, Zr, Al, Fe, and B, while R is typically an alkyl group (C_xH_{2x+1}) [4].

The sol-gel process involves a transition from a liquid colloidal solution (sol) to a solid three-dimensional matrix (gel). The precursors for the synthesis of these colloids consist of a metal or metalloid element surrounded by several reactive ligands. Metal alkoxides are the most popular due to their affinity with water. Among the most used alkoxylated agents are tetramethoxysilane (TMOS), tetraethoxysilane (TEOS), and alkoxides such as aluminates, titanates, and borates mixed with TEOS [1].

In the hydrolysis reactions (**Figure 1(1)**), alkoxide groups (—OR) are replaced with hydroxyl groups (—OH) through the addition of water. Subsequent condensation reactions (**Figure 1(2a)** and **(2b)**) involve the silanol groups (Si—OH), which create siloxane linkages (Si—O—Si) and subproducts such as water and alcohol. The condensation reaction starts before the hydrolysis has been completed. However, conditions such as the pH, H_2O/Si R and catalyst can force the hydrolysis to end before the condensation starts [3].

The use of an alcohol favors hydrolysis, causing the miscibility of the alkoxide and the water [3]. As the number of siloxane linkages increases, the individual molecules are joined to one another and aggregated into the sol. When the sol's particles are interlocked into a network, a gel is formed. A drying stage in which volatile components such as water and alcohol are extracted from the network is necessary, causing the gel to shrink while condensation occurs. It should be noted, however, that the addition of solvents and certain reaction conditions can promote esterification and depolymerization reactions, in accordance with the reverse reactions

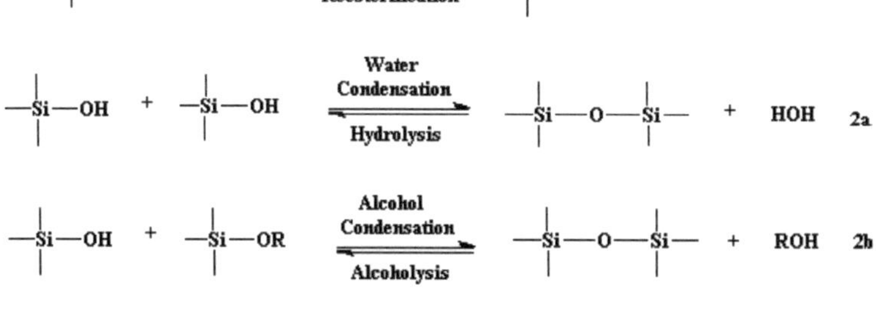

Figure 1.
General schematic of the hydrolysis and condensations of alkoxides.

(**Figure 1(2a)** and **(2b)**) [1, 3]. More specifically, factors such as the pH, nature and concentration of the catalyst, and H_2O/Si R play the most important parts in the final structure and properties of the obtained hybrid polymeric network [1]. In this section, the effect of pH will be described, as controlling the pH of sol-gel media is generally used to obtain either coatings or powders; however, readers can find a detailed mechanism in specialized reviews or books on sol-gel chemistry.

Drying involves the loss of water, alcohol, and other volatile components. During drying, the gel initially shrinks due to the loss of pore fluid that maintains the liquid-vapor interface on the outer surface of the gel, this occurs always under atmospheric conditions. The liquid-vapor meniscus recedes toward the interior of the gel, in the final stage of drying [5]. When the drying is under supercritical conditions, the surface tension disappears with a gradient of capillary pressure accumulated in the walls of the pores, which avoids the possible collapse of the volume of the pores due to the capillary forces. Under these conditions, the materials are left with a wet porous texture that prevents the collapse of the pores of the gel, and the resulting materials are generally hydrophobic, because their surfaces are covered with alkoxy groups [6].

2.1 Sol-gel coatings

For anticorrosion and biomedical applications, several techniques are used to surface modify metals to improve their mechanical properties, enhance their corrosion resistance or, in some cases, give biological, osteoconductive, or antibacterial activity. Among these techniques are physical vapor deposition (PVD), chemical vapor deposition (CVD), and electrophoresis, and sol-gels can be used to surface modify metal [7]. However, the versatile methodology of sol-gel synthesis generates diverse types of materials that have found applications in several scientific and technical fields [8]. It is precisely this versatility that has generated great interest in using sol-gel techniques to develop coatings that are applied in areas such as analytical chemistry to develop more efficient, specific sorbents that allow the concentration of desired analytes [9–11]. In biomedical fields, sol-gel coatings have gained attention in controlling the surface interactions between medical implants/devices and biological environments [12, 13]. In the field of photocatalysis, sol-gel coatings have been developed for applications such as for organic compound degradation [14]. Anticorrosion sol-gel coatings have been applied to avoid degradation of materials and metallic structures to prevent or preserve the surface and bulk integrity of metallic materials [15].

The sol-gel coating technique consists of the immersion of a substrate that is to be coated in the "sol" solution and the vertical extraction of this substrate at a controlled speed [13]. A very fine coating of gel is thus formed, since there is rapid evaporation of the solvent during the extraction of the substrate. The thickness depends on the viscosity of the liquid, the surface tension, and especially the speed of removal; the higher the speed of removal is, the greater the thickness of the coating [16]. Once the first coating layer is obtained, the process can be repeated to form a multilayer structure. The drying step also influences the final structure of the film, and there are thickness limits that must be obeyed to avoid the cracking of films or their detachment from the substrate [17].

3. Sol-gel coatings for anticorrosion purposes

According to the standard ISO8044:2015, corrosion is defined as a spontaneous degradation of metals due to their physicochemical interaction with the surrounding environment, which changes the properties of the metal and can lead to its functional destruction. The main cause of the corrosion of metals is their thermodynamic instability in outdoor conditions. As a result, metals are converted into compounds (oxides, hydroxides, carbonates, sulfides, etc.) that, as corrosion products, are the most stable forms of the metal. In this way, corrosion leads to a decrease in Gibbs free energy and therefore occurs spontaneously. Generally, coatings are designed to stop corrosion of metals by one of the following mechanisms: cathodic protection, anodic passivation, electrolytic inhibition, environmental modification, plating, painting, and active corrosion inhibition [18, 19].

Using coatings to protect metallic substrates from corrosion is an active and important research area in materials science and industry. Through the application of coatings, corrosion can be minimized and controlled; the coating acts as a barrier preventing contact between the corrosive medium and the metallic substrate and preventing ion migration among the coatings; in addition, in cathodic protection, the coating material acts as a sacrificial anode. The use of species for inhibition/passivation, including cases of anodic and/or cathodic protection, inhibits the action of external corrosive agents. The sol-gel process stands out among the many other coating methods, e.g., CVD [20], PVD [21], electrochemical deposition [22], plasma spraying, and others [23]. The sol-gel coating process generally involves temperatures close to room temperature; thus, thermal volatilization and degradation of entrapped species, such as organic inhibitors, is minimized. Since liquid precursors are used, it is possible to cast coatings in complex shapes and to produce thin films without the need for machining or melting. The sol-gel films are formed by "green" coating technologies, which use compounds that do not introduce impurities into the product; this method is waste free and lacks a washing stage. The interest in this type of material has increased exponentially in recent decades.

3.1 Organic-inorganic hybrid coatings

Organic-inorganic hybrid (OIH) coatings created by the sol-gel process are very suitable for resisting corrosion. Inorganic sols in hybrid coatings not only increase adhesion by forming chemical bonds between metals and hybrid coatings but also improve the comprehensive performance of polymers in the coatings. Different organic polymers or organic functionalities are introduced into the gel network to produce tailored properties such as hydrophobicity and increased crosslink density. For corrosion protection of metals, organic components of hybrid coatings are selected to repel water, form dense thick films, and reduce coating porosity. Factors such as the ratio between inorganic and organic components in hybrid coatings, cure temperature, and pigment-related parameters need to be optimized as a function of the specific metal for the production of hybrid films with maximum corrosion resistance [24].

Sol-gel OIH coatings are macromolecular matrices where the intermolecular interactions (such as porosity, rigidity and adhesion to the substrate, among others) between its structure and the metal surface are very relevant for the final properties of the material. The better it is combination between OIH coatings and the substrate results in materials with improved protection against corrosion, oxidation and erosion, and good electrical and thermal insulation properties. Sol-gel OIH coatings are commonly produced by gels obtained from the gelation of colloidal solutions, hydrolysis and polycondensation of precursors, and their subsequent aging and drying [25].

Sol-gel OIHs based on siloxanes (Si—O-metal oxide), alkoxysilanes and alkoxides of zirconium, titanium, cerium, tin, and aluminum are potential candidates for the treatment of steel surfaces, allowing covalent bonding between the inorganic parts of the OIHs and the metallic substrate.

For iron-based alloys, steel is the most common and versatile metal, and the corrosion resistance behavior of sol-gel coatings deposited on these substrates has been extensively studied. Publications have mostly focused on the use of TEOS, 3-glycidoxypropyl-trimethoxysilane (GPTMS), methacryloxypropyl trimethoxysilane (MAPTS), and methyltriethoxysilane (MTES) and have sometimes included the performance of coating materials with embedded corrosion inhibitor species. For example, Santana et al. used clay nanoparticles to improve mechanical and barrier properties and Ce as an inhibitor for OIH sol-gel coatings [26]. Generally, the reported studies show that those OIH coatings exhibit a promising performance against corrosion in iron-based alloys [27]. The use of different organosilica sol-gel coatings on steel substrates has been reported. The results indicate that it is possible to tailor the sol-gel composition to modify coating properties such as hydrophobicity, wettability, adhesion, and corrosion prevention. Precise selection of organically modified sol-gel compositions has yielded coatings that are adhesive, water-repellent, and effective at preventing corrosion of coated steel panels.

OIH sol-gel coatings are an excellent choice for aluminum-based alloys because in addition to the OIH properties; these materials can provide a stable Si—O—Al bond between the inorganic functionality of these materials and the formed passivation layer. Previous work in the literature focuses on the use of TEOS and GPTMS, the performance of coating materials with encrusted corrosion inhibitor species, and the deposition of multilayers by several cycles of deposition curing [25, 28]. Cambon et al. investigated modifications of OIH sol-gel coatings using different amounts of cerium, studied how these coatings protected different aluminum-based alloys from corrosion with electrochemical methods, and reported improvements in the anticorrosion process by increasing the concentration of cerium in OIH sol-gel coatings [29].

For copper and copper-based alloys, the use of OIH coatings did not give very good results with respect to corrosion, but the use of TEOS and GPTMS with 1,2,3-benzotriazole resulted in a corrosion inhibitor effective in different environmental media. For zinc-based alloys, few studies have used OIH sol-gel coatings, and GPTMS, TEOS and MTES have been used as precursors to focus on new green conversion coatings based on molybdate, permanganate, silicate, titanate, rare earth salt, tungsten, and vanadate compounds. Finally, for magnesium-based alloys with OIH coatings, the reported literature mainly focuses on the use of GPTMS in combination with other reagents, especially (3-aminopropyl) triethoxysilane (APTES). For this material, the inorganic component is selected to form the network for the film, while the organic component is selected to repel water and fill the porosity shortly. These hybrid coatings have excellent mechanical strength and adhesion to metal substrates. Hybrid coatings doped with slow release corrosion inhibitors provide long-term metal anticorrosion. Superhydrophobic coatings are an excellent option to resist corrosion, and their properties are derived from the low surface tension and roughness of the surface of hybrid coatings, although it is necessary to prolong their durability. It is obvious that the combination of these techniques can provide superior anticorrosive properties [24].

Despite the advantages of combining different properties in these materials, synthesis constraints remain. The major limitations of sol-gel processing for coating metals are delamination, crackability, adhesion, and thickness limits. Assuring a uniform distribution on the substrate and optimizing thermal treatments (curing/drying) are crucial factors to ensure the quality of anticorrosive coatings.

3.2 Smart coatings

Active corrosion inhibition addresses the unavoidable failure of coatings and includes the introduction of components that release selectively during damage to the coating to reconstruct a protective barrier at the metal-environment interface. Such active corrosion inhibition is different from the broader concept of "self-healing," which also includes the introduction of materials that are released within the coating that allow for reformation of polymeric organic coatings even without direct protection from corrosion [18].

Functional coatings (organic, inorganic, or hybrid) are a new class of materials that can be adapted for many applications in which they should be able to perform well-defined arrays of functions. The concept of a smart coating is more recent and has been applied to functional coatings that can respond to certain stimuli generated by surroundings [30]. Functional and smart coatings have been regularly reviewed for various applications, including active corrosion protection and inhibition. However, summarizing the progress in this area requires a concise review of the latest trends. The application of functional and smart coatings is one of the most promising routes to developing active corrosion protection and inhibition systems. Two main strategies have been pursued to introduce the required functionalities into coatings: encapsulation/loading of functional active species in host carriers and manipulation of the coating matrix composition for inclusion of bulk and/or surface functional groups. In this case, intelligent self-healing coatings are coatings whose suitable repair agents are safely stored but can be released on demand, i.e., when corrosion occurs and only then [31].

3.3 Self-healing coatings by sol-gel methods

Self-healing materials are well known as materials that are capable of autonomously restoring their properties in such a manner that they can function longer than similar materials without self-healing abilities. Self-healing coatings are required for the total or partial repair of coated areas damaged by aging or unexpected aggressive events. Two main strategies have been pursued regarding self-healing coatings for corrosion protection: (i) mending defects formed in the polymeric coating matrix via addition of polymerizable agents and (ii) inhibition of corroding areas due to the presence of corrosion inhibitors [32, 33]. Usually, the self-healing agent is released because of mechanical damage. However, not all mechanical damage leads to corrosion, and corrosion is not necessarily initiated at mechanical cracks. Ideally, the release of self-healing agents should take place only when corrosion is initiated. Triggers for sensing the corrosion of a metal system that have been investigated in depth are pH and ionic strength changes. However, the most reliable and case-selective trigger is a change in the electrochemical potential, as it always decreases when corrosion occurs [31].

Currently, the literature exhibits many reports of the encapsulation of agents for protection against corrosion that have self-healing capacity activated by different extrinsic or intrinsic stimuli, which can be local pH gradients, capsule rupture induced by mechanical stress, ion exchange processes, water trapping, electrochemical potential changes, light irradiation, thermal variations, and others; in some cases, neither the kinetics of healing nor the mechanisms are completely understood [30]. Below, some self-healing coatings for corrosion protection that were obtained by the sol-gel technique and have been applied on aluminum substrates are presented.

Snihirova and coworkers investigated inhibitor-doped hydroxyapatite (HAp) microparticles that were used as reservoirs, storing corrosion inhibitors to be released on demand. Release of the entrapped inhibitor was triggered by redox reactions associated with the corrosion process. HAp was used as reservoirs for several inhibiting species, Ce(III), La(III), salicylaldoxime, and 8-hydroxyquinoline, which are effective corrosion inhibitors for 2024 aluminum alloy (AA2024).

Dissolution of the microparticles and release of the inhibitor were triggered by local acidification resulting from an anodic half-reaction during corrosion of AA2024. Calculated and experimentally measured local acidification at the aluminum anode (down to pH = 3.65) were presented. The anticorrosion properties of inhibitor-doped HAp were assessed using electrochemical impedance spectroscopy. Microparticles impregnated with corrosion inhibitors were introduced into a hybrid silica-zirconia sol-gel film, which acted as a thin protective coating for AA2024, an alloy used for aeronautical applications. The protective properties of the sol-gel films were improved by the addition of Hap microparticles, proving their applicability as submicrometer-sized reservoirs of corrosion inhibitors for active anticorrosion coatings. The synthesis of HAp from solutions of calcium nitrate and ammonium hydrophosphate was carried out with the addition of citric acid to modulate its morphology. The obtained solution was placed in a water bath at 37°C for 24 h, allowing the precipitation of HAp. Once the HAp particles were obtained, they were immersed in solutions containing Ce(III), La(III), salicylaldoxime, and 8-hydroxyquinoline. Hybrid silica-zirconia sol-gel films were obtained by combining an organosiloxane alkosol with another alkosol containing a zirconia. The silane-based alkosol was prepared through hydrolysis of GPTMS in 2-propanol by adding a diluted aqueous solution of HNO_3 at room temperature for 1 h. The second alkosol, containing zirconia nanoparticles, was prepared through controlled stoichiometric acidic hydrolysis of a 70 wt% 2-propanol solution of Zr(IV) tetrapropoxide in ethyl acetoacetate under ultrasonic agitation. Finally, the two alkosols were mixed in a 2:1 volume ratio, ultrasonically agitated for 1 h and then aged for another 1 h at room temperature [34]. According to their results, HAp presents a good chemical stability and compatibility with the sol-gel matrix, sufficient loading capacity, an ability to sense corrosion onset (local acidification), and an ability to release the inhibitor on demand [34].

Li and coworkers have reported the incorporation of environment-responsive properties into tube-like nanomaterials in self-healing coatings for corrosion protection. Stimulus-responsive silica/polymer double-walled hybrid nanotubes with a controlled aspect ratio (length/diameter) were synthesized by surface-graft precipitation polymerization. The surface grafts on the hybrid nanotubes consisted of pH-, temperature-, or redox-responsive polymers that can confer a smart stimulus-responsive property upon the hybrid nanotubes. In addition to their well-defined morphology, uniform size, and wall thickness, the as-prepared silica/polymer hybrid nanotubes exhibited release in response to these different environmental stimuli. The silica/polymer double-walled hybrid nanotubes serve as intelligent nanocontainers of the anticorrosion agent benzotriazole.

The silica/polymer double-walled hybrid nanotubes served as smart nanocontainers, which is very important for applications in self-healing coatings [35]. Self-healing coatings were prepared by dispersing the as-obtained nanotube containers into SiOx/ZrOy hybrid films at room temperature. The SiOx/ZrOy hybrid films doped with benzotriazole-loaded SiO_2/PMAA nanocontainers were prepared by a sol-gel process. In this case, a zirconium oxide sol was prepared from a tetra-n-propoxy zirconium (TPOZ) solution (70 wt% 2-propanol) in ethylacetoacetate at room temperature. The second organosiloxane sol was prepared by hydrolyzing 3-chloropropyltrimethoxysilane (CPTMS) in 2-propanol by the addition of acidified water (HNO_3, pH 0.5). Then, the zirconia sol and organosiloxane sol were mixed, stirred and aged at room temperature. BTA-loaded nanotube containers were mixed with the SiOx/ZrOy films at a concentration of 10 mg/mL by sonication. Self-healing films were fabricated on carbon steel by a dip-coating procedure. The carbon steel was immersed into the sol-gel mixtures and then withdrawn, and the samples were cured at 130°C [35].

As you will see in detail later, corrosion is an important factor in the design and selection of alloys for application in vivo. Because toxic species might be released to the body during corrosion and various corrosion mechanisms can lead to implant loosening and failure, biomaterials are often required to be tested for corrosion and solubility before they are approved by regulatory organizations.

4. Sol-gel coatings applied to biomaterials

A biomaterial is any substance that has been engineered to interact with biological systems for a medical purpose, such as a therapeutic (which treats, augments, repairs, or replaces a tissue function of the body) [36–38]. Metals [23], polymers [39], ceramics [40], and their combinations [41] can be used as biomaterials. Among the different kinds of biomaterials, metal implants have been used with success in orthopedic applications, as well as temporary appliances such as pins, screws, and bone plates and permanent implants such as total joint substitutions.

An implanted material could have a desired or undesired biological response depending on its surface characteristics such as wettability, charge, topography, and surface chemistry [42]. These surface properties are the major factors that ultimately determine the rejection or acceptance of a biomaterial in the body because the surface is the interface where the biomaterials meet and interact with the biological environment (i.e., bone, soft tissue, and blood). For this reason, surfaces of these metals must be modified; otherwise, the body will recognize those implants as foreign and isolate them in a fibrotic capsule, compromising their overall performance [43]. Surface modification of these metallic surfaces is engineered and designed to improve tissue tolerance, osseointegration, and corrosion [44]. Among the different surface modification technologies, such as grist blast [45], chemical etching [46], and plasma surface modification [47], the application of sol-gels has also been shown to improve the biocompatibility of metal implants [48].

Osseointegration is defined as “the direct structural and functional connection between the ordered, living bone and the surface of a load-carrying implant” [49]. The sol-gel process allows the chemical composition of its products to be controlled, resulting in greater bioactivity than those materials with the same composition but prepared with different methods [13]. The intrinsic bioactivity of sol-gel materials together with their possibility of being coupled with a range of coating techniques, e.g., dip, spin, and spray coating, makes sol-gels an ideal technology in the making of bioactive and biocompatible coatings [13].

In this section, TiO_2, HAp, bioactive glasses, and hybrid composite coatings synthetized by the sol-gel technique to surface-modify metals of biomedical interest for implants will be described.

4.1 Hydroxyapatite (HAp) as a sol-gel coating

HAp ($Ca_{10}(PO_4)_6(OH)_2$) is a biocompatible and bioactive material that can be used to restore bony defects [50–52]. Because of its chemical and structural similarities to the inorganic phase of human bone, HAp shows excellent biocompatibility. Most of the relevant applications of this material take advantage of its biological influence on tissues and especially its biodegradation behavior. Conventionally, HAp powders are synthesized via precipitation from an aqueous solution, solid-state reactions, and hydrothermal methods. However, these powders are sometimes unsuitable for the preparation of HAp ceramics with controlled biocompatibility and bioactivity because of their chemical and phase heterogeneity [53–58].

One of the most important issues when considering HAp for biological coatings is its dissolution rate in human body fluids. Plasma-sprayed HAp coatings dissolve and degrade quickly, resulting in the weakening of the coating-substrate bonding or implant fixation to host tissues [59, 60].

Sol-gel synthesis of HAp ceramics has recently attracted much attention because of its advantages: it provides a molecular-level mixing of the calcium and phosphorus precursors and is capable of improving the chemical homogeneity of the resulting mineral composite in comparison with the products from conventional solid-state reactions, wet precipitation, and hydrothermal synthesis [55–57, 61, 62].

The sol-gel process provides an alternative to conventional synthesis methods. The metal alkoxides, $M(OR)_n$, convert to amorphous gels of metal oxides through hydrolysis and condensation reactions. The sol-gel materials are transformed into ceramics by heating at relatively low temperatures and have better chemical and structural homogeneity than ceramics obtained by conventional methods.

Pathway alkoxide-based considers an anhydrous medium, where precipitation of unwanted calcium phosphate phases can be slowed down, and the molar concentration of the reactive species can be heightened relative to a point that benefit the mineral development before the start of the reaction. Likewise, chelation-based sol-gel methods preserve a good degree of control, preventing the precipitation.

The synthesis of pure HAp requires the same molar ratio between calcium and phosphate as in the final product. The temperature required to form the apatite structure depends on the chemical nature of the substrate. Low-temperature synthesis particularly benefits the metal substrates (temperatures below 500°C).

The conditions to obtain the best coatings have been related to the pH decrease that occurs in aqueous sols during aging, which is based on the polymerization reaction between calcium and phosphorous. To obtain homogeneous, crack-free coatings, the annealing temperature and thickness of the coatings should also be controlled. Film roughness is related to the viscosity of the sol-precursor used for deposition and the number of coating layers. Thus, depending on the exact application, various parameters of the applied coating can be adequately controlled.

Prepared HAp sols and Ag-doped HAp sols have been coated on passivated Ti surfaces by spin coating at 5000 rpm for 50 s. The coated Ti surfaces were immediately dried at 70°C for 12 h and then heat-treated at 650°C for 3 h. HAp without Ag doping was used as controls in this study. All samples were autoclaved prior to material characterization and all culture experiments [63].

Overall, it was concluded that the material AgHA1.0 has similar biological activity as HAp with respect to bone cell proliferation and differentiation. In addition, AgHAp 1.0 also minimizes the initial bacteria adhesion [63].

Caution should also be taken in applying HAp-coated materials to patients, since an HAp coating was found to have negative effects on bone formation. Reports in the literature [64, 65] have suggested that HAp coatings are unstable, susceptible to bacterial infection, possibly predisposed to rapid osseous breakdown, and absent of significant advantages over titanium implants [66].

To obtain bioactive materials with high mechanical strength, metal implants (titanium and stainless steel (SS) alloys) are usually coated with a thin layer of HAp using plasma spray techniques [67–69]. The main problem associated with this technique is a lack of exact stoichiometry control and the occurrence of glassy phases in the ceramic layer. Some of these additional phases do not show bioactive behavior or do not dissolve in biological environments [69].

For the above, various techniques have been described as effective means of surface treatment and corrosion resistance enhancement of NiTi implants, including chemical passivation, electropolishing, anodization, thermal oxidation, laser surface melting, nitriding, plasma ion implantation, and coating with sol-gels [70–75].

Out of these, the sol-gel route for surface modification of Ti implants is of particular interest because of simple and inexpensive methodology, low temperature processing, and suitability for coating substrates of irregular shapes, such as implants [76–78].

4.2 Titanium dioxide (TiO_2) as a sol-gel coating

TiO_2 is a very interesting material due to its technological applications in biomaterials. An example of this is sol-gel-derived titania and its coatings, which are bioactive materials, and are used in applications as diverse as biomaterials [79, 80]. Advantages of sol-gel processes over other methods are their controlled transformation of the microstructure of the deposited film. Today microstructure control is needed for many applications and the use of sol-gel routes opens up new possibilities.

In common sol-gel methods for preparing TiO_2 materials, highly reactive alkoxide titanium precursors are violently hydrolyzed and further condense to form a Ti—O—Ti network. Unfortunately, this route can lead to precipitation of amorphous particles with uncontrolled structures. To overcome many of the specific problems of sol-gel methods employing water as the hydrolysis agent and to control hydrolysis and polycondensation reactions, the application of nonhydrolytic methods, ionic liquids, organic additives, and coordination chemistry have been attempted [81–83].

On the other hand, it has been reported that TiO_2 films can effectively improve corrosion resistance and biocompatibility [84–86]. An example of this is given in titania films on biomedical steel exhibit anticorrosion properties in physiological solutions [87, 88]. At present, resistant sol-gel TiO_2 coatings synthesized by heat treatment under low temperature have received good results for the preparation of anticorrosive and bioactive coatings to protect metal implants.

The desired properties can be obtained by careful control of reaction conditions or by the use of suitable additives [89]. Moreover, sol-gel processing offers a unique opportunity to prepare layers at low temperatures so that an essential part of the sol-gel preparation process is the thermal treatment that is necessary to form pure TiO_2 films.

Conventional metal implants can be surface modified with sol-gel coatings, but organic synthetic polymers such as polyether-ether-ketone (PEEK) can also be surface modified with TiO_2 sol-gel-derived coatings to overcome low bioactivity due to their chemical inertness, which is characteristic of most synthetic organic polymers. These sol-gel-derived TiO_2 coatings showed greater bone bonding ability than PEEK [90].

However, to avoid corrosion, a TiO_2 dip-coating method and its variants are commonly used for the deposition of TiO_2 sol-gel-derived coatings onto metal implants [91]. For example, an NiTi surgical alloy was surface modified with thin films of TiO_2 to improve corrosion resistance, but these films also showed blood compatibility in vitro [92]. Avoiding corrosion and improving biocompatibility with surrounding tissues are not the only ultimate goals of sol-gel-coating metals of biomedical interest, avoiding the release of toxic ions, possibly through degradation, that some alloys may contain is another such goal [92].

With respect to annealing time and temperature in sol-gel coatings, both parameters exhibit a dependence with the degradation rate, as studied in the TiO_2 sol-gel deposited onto a magnesium alloy (AZ31), where the treatment with low annealing temperatures decreased the corrosion rate. Long-time treatment of annealing helped to enhance corrosion resistance [93].

TiO_2 sol-gel coatings doped with a high calcium ion concentration showed better corrosion resistance for M30NW biomedical alloy substrates in a simulated body fluid (SBF) test than similar coatings with low calcium ions concentrations [94].

4.3 Sol-gel-derived bioactive glass coatings

Bioglasses are a family of materials that have shown bioactivity for bone repair and can bond with living bone [95]. In 1971, the first bioglass named 45S5 was discovered by Hench, and since then, many other glass compositions have been developed. Bioglass 45S5 is composed of 45 wt% SiO_2, 24.5 wt% CaO, 24.5 wt% Na_2O, and 6.0 wt% P_2O_5, but other similar composition has been used and in some cases enhancing components can be added [96]. Bioglass 45S5 compositions have been shown to be optimal for biomedical applications because it is similar to that of HAp, the mineral component of bone. Ca/P ratios in $SiO_2{\cdot}CaO{\cdot}P_2O_5$ glasses coatings can be controlled with stoichiometric control of TEOS, calcium nitrate tetrahydrate ($Ca(NO_3)_2{\cdot}4H_2O$), and triethyl phosphate (TEP) as sol-gel precursors [97]. Coatings with higher Ca/P ratios showed that greater cell proliferation, however, growth inhibition was observed in response to a low Ca/P-ratio in coating compositions [98].

The surface of a bioglass implant, when subjected to an aqueous solution or body fluids, converts to a silica-CaO/P_2O_5-rich gel layer that subsequently mineralizes into hydroxycarbonate in hours [99–101]. Bone tissue growth improved with increasing dissolution [102]. This gel layer resembles the HAp matrix so much that osteoblasts were differentiated and new bone was deposited [103].

Bioglasses are the most promising materials for bone grafting in several clinical applications such as orthopedic, dental, maxillofacial, and otolaryngological applications [104].

From a synthetic point of view, bioglasses can be prepared either by melting or sol-gel methods, which affect their physical and biological properties [105], but it is also important to define the methods of coating preparation or deposition affecting the ultimate performance of the coatings. Bioglass coatings are usually deposited onto metals or alloys using sol-gel [106], electrophoretic deposition [107], laser cladding [108], and thermal spraying (plasma spraying and high-velocity oxy-fuel) techniques [109]. The technique most employed to spray bioglass since 1980 is thermal spraying and specifically atmospheric plasma spraying, due to its low cost and industrial feasibility [110, 111].

Sol-gel-derived bioglasses are excellent materials for use in tissue engineering applications, such as covering prosthetic metallic implants. Recently, porous bioactive glasses have been derived through sol-gel processing in an attempt to increase the specific surface area and thus the surface reactivity and degradability of the material. This approach allows the material to be replaced ultimately by natural tissue while stimulating bone regeneration [97].

The sol-gel technique can be used to coat 316 L SS [112], titanium [4], and magnesium biomedical alloys [106] with bioactive glass or derived glass-ceramic.

For example, the formation of an apatite layer assures the bioactivity of the bioglass coating, which also improves the corrosion resistance of 316 L SS substrates. Bioactive glass-coated 316 L SS showed greater pitting corrosion resistance than pristine samples. It was concluded that by using the bioactive glass-coated 316 L SS as a human body implant, improvement of corrosion resistance, as an indication of biocompatibility, and bone bonding could be obtained simultaneously [112]. Uncoated 316 L SS possesses high corrosion current density (Icorr = 265 nA/cm^2) and thus low corrosion resistance in normal saline solution [112].

Relatively dense sol-gel coatings can be obtained with postheat treatment, causing a substantial volume contraction. In parallel, residual stress gradually accumulates at the interface between the coating and the substrate [113], remarkably affecting bond strength and the corrosion resistance of samples [114]. Therefore, heat treatments must be carefully controlled and optimized to obtain favorable bonding properties as well as corrosion resistance for sol-gel-derived coatings on magnesium alloy implants [106].

The sol-gel process is carried out at much lower temperatures than traditional melting methods. Because of the low fabrication temperatures used in this method, the composition and homogeneity of the product are greatly controlled. Higher mesoporosity and surface areas are obtained in sol-gel-derived bioglasses than in melt-derived bioglasses, which exhibit high glass transition temperatures [105].

4.4 Organic-inorganic composite hybrid coatings

Inorganic sol-gel coatings are brittle, which can compromise their performance. To overcome this drawback, an organic polymer can be entrapped in an inorganic sol-gel glassy matrix to form an OIH sol-gel nanocomposite coating that can be deposited onto different metals and their alloys of biomedical interest. Poly-ε-caprolactone (PCL) [13], poly(DL-lactic-glycolic acid) [115], silica-polyethylene glycol hybrids [116], chitosan [117], and collagen type I have been incorporated into sol-gel coatings.

Sol-gel coatings that can release silicon compounds under in vivo conditions have been shown to promote fast and good osseointegration. For this reason, hybrid composites can be prepared through acid-catalyzed sol-gel methods using methyltrimethoxysilane (MTMOS) and GPTMS as alkoxide precursors, which allows the degradation kinetics and Si release of the coatings to be controlled by adjusting the amount of GPTMS. Although these coatings showed osteoinduction ability in vivo, coatings with some alkoxide proportions did not demonstrate strong cellular results [118].

Another way to control the degradation profiles of HAp coatings is to incorporate silver ions, which are effective in inhibiting microbial infection [119].

For example, type I collagen layers can be assembled into a sol-gel composite coating to cover magnesium alloys such as AZ31 and ZE41 to allow the release of growing factors, enhancing cell adhesion for tissue integration. This effect is due to the high biocompatibility and cytocompatibility that type I collagen has as well as the its positive effects on cell activity [120].

Another bioactive composite coating, one composed of a silica xerogel and chitosan hybrid, was used to surface modify Ti at room temperature through a sol-gel process to obtain crack-free thin layers (<2 μm) with a chitosan content of >30 vol.%. These hybrid coatings showed bioactivity, and their properties suggested applicability to titanium-based medical implants [121].

From a biological perspective, titanium is classified as a biologically inert material that does not promote adverse reactions and is well tolerated by human tissues. However, the formation of peri-implant fibrosis may isolate the implant from the surrounding bone and induce the mobilization of prostheses, thus reducing their performance [122].

Surface hydroxy groups enhance the bioactivity of sol-gel glasses due to their promotion of calcium phosphate deposit nucleation, causing osseointegration when these materials are implanted [76].

Commercially pure (CP) Ti grade 4 substrates were dip coated with an OIH crack-free coating consisting of a sol-gel-derived ZrO_2–based matrix in which different PCL percentages (between 20 and 30 wt%) were incorporated. These films showed bioactivity and induced HAp formation when they were soaked in SBF. Biological evaluation with human mesenchymal stem cells (hMSCs) demonstrated that compared to pristine Ti, the coatings were nontoxic, supported cell proliferation at all compositions, and did not hamper hMSC differentiation in an osteogenic medium [122].

Although further investigation is needed to fully describe the osseointegration potential of the developed OIH, such a material may find application in the surface coatings of Ti dental implants [122].

The addition of an organic polymer such as PEG onto SiO_2 in a hydric composite can affect its biocompatibility and bioactivity, as cell growth and the proliferation of NIH 3T3 cells depend on the PEG amount and exposure time. The formation of a HAp layer was indeed observed on the material's surface by scanning electron microscopy (SEM)/energy-dispersive X-ray (EDX) analysis after the samples were soaked in SBF. Their effects were superior to those exerted from SiO_2, whose ceramics are already widely used in medical applications [123].

Homogeneous, porous and crack-free titania-based coatings can be obtained when PCL is added to a TiO_2 inorganic matrix to make pure grade 4 titanium (CP Ti-4) disks more bioactive, as demonstrated by the ability to induce the formation of HAp when soaked in SBF, which is a crucial property for the osseointegration of metal implants in vivo [13].

Hybrid inorganic-organic ZrO_2/PCL composite films showed HAp formation on their surface when the hybrid coating was exposed to SBF, implying their osteointegration ability once implanted in vivo. In addition, a WST-8 colorimetric assay shows that the coating makes Ti-4, which is generally bioinert and biocompatible [124].

Finally, it has been demonstrated that biodegradability can be controlled by adjusting the composite sol-gel hybrid coating composition, resulting in the ultimate material performance.

5. Conclusions

Sol-gel coatings provide excellent corrosion protection by providing a protective barrier layer for the reduced permeation of corrosive entities, providing a water repellent surface, chemically modifying the surface of a metal to make it more inert, and altering the electrical potential of surface sites. This review has presented an overview of new approaches to generate self-healing behavior in smart coatings. Extending the working life of structural and industrial metallic surfaces might depend on the fabrication of novel self-healing protective smart coatings that are able to repair scratches and eliminate corrosion. Additional development of materials with self-healing properties will reduce the loss that results from corrosion of metallic materials that are used in chemical plants, automobile parts, building structures, and home appliances.

The factors most limiting the use of sol-gel processing for coating metals are delamination, crackability, adhesion, and thickness limits. Assuring a uniform distribution on the substrate and optimizing thermal treatments (curing/drying) are crucial factors in ensuring the quality of anticorrosive coatings.

Although other processes such as plasma spraying, CVD, and electrochemical methods can be used to obtain thin film coatings of TiO_2, HAp or bioglass on metallic substrates for biomedical purposes, the sol-gel process has remarkable advantages over those techniques, including better control of composition, structures, and porosity, which results in greater bioactivity than materials with the same composition but prepared with other techniques. Sol-gel coatings, due to their low processing temperatures, can also be applied onto nonmetallic implant substrates, such as organic polymers such as PEEK, or nonpermanent metallic implants such as those made of magnesium alloys. In addition, the sol-gel process can perfectly enable the integration of organic polymers with an inorganic glassy matrix.

Author details

Cristian Carrera-Figueiras, Yamile Pérez-Padilla, Manuel Alejandro Estrella-Gutiérrez, Erbin G. Uc-Cayetano, Juan Antonio Juárez-Moreno and Alejandro Avila-Ortega*
Facultad de Ingeniería Química, Universidad Autónoma de Yucatán (UADY), Mérida, Yucatán, Mexico

*Address all correspondence to: alejandro.avila@correo.uady.mx

References

[1] Brinker CJ, Scherer GW. Sol-Gel Science, the Physiscs and Chemistry of Sol-Gel Processing, Advanced Materials. Vol. 3. Deerfield Beach, Fla: Academic Press; 1990. 1-908 p

[2] Levy D, Zayat M. The Sol-Gel Handbook. David Levy, Zayat M, editors. Vols. 1-3, USA: Wiley-VCH; 2015. 1-1508 p

[3] Schaefer DW, Keefer KD, Brinker CJ. Structure of silica condensation polymers. In: Zeigler JM, FWG F, editors. American Chemical Society, Polymer Preprints, Division of Polymer Chemistry, Advances in Chemistry. Vol. 24. Washington, DC: American Chemical Society; 1983. pp. 239-240

[4] Durán A, Conde A, Coedo AG, Dorado T, García C, Ceré S. Sol–gel coatings for protection and bioactivation of metals used in orthopaedic devices. Journal of Materials Chemistry. 2004;**14**(14):2282-2290

[5] Smitha S, Shajesh P, Aravind PR, Kumar SR, Pillai PK, Warrier KGK. Effect of aging time and concentration of aging solution on the porosity characteristics of subcritically dried silica aerogels. Microporous and Mesoporous Materials. 2006;**91**(1-3):286-292

[6] Pierre AC, Pajonk GM. Chemistry of aerogels and their applications. Chemical Reviews. 2002;**102**(11):4243-4266

[7] Guglielmi M. Sol-gel coatings on metals. Journal of Sol-Gel Science and Technology. 1997;**8**(1/2/3):443-449

[8] Wright JD. NAJMS. Sol-gel materials: Chemistry and applications. 1st Edition. John D. Wright NAJMS, editor. London: CRC Press; 2001. 136 p

[9] Chong SL, Wang D, Hayes JD, Wilhite BW, Malik A, Chong SL, et al. Sol-gel coating technology for the preparation of solid-phase microextraction fibers of enhanced thermal stability. Analytical Chemistry. 1997;**69**(19):3889-3898

[10] Jillani SMS, Alhooshani K. Urea functionalized surface-bonded sol-gel coating for on-line hyphenation of capillary microextraction with high-performance liquid chromatography. Journal of Chromatography. A. 2018;**1543**:14-22

[11] Burgos-Tan MJ, Pérez-Padilla Y, Avila-Ortega A, Barrón-Zambrano JA, Vilchis-Néstor AR, Carrera-Figueiras C, et al. Preparation, characterization and evaluation of a hybrid polymeric coating with sorbent properties. Chemical Papers. 2017;**71**(7):1205-1215

[12] Owens GJ, Singh RK, Foroutan F, Alqaysi M, Han C-M, Mahapatra C, et al. Sol-gel based materials for biomedical applications. Progress in Materials Science. 2016;**77**:1-79

[13] Catauro M, Papale F, Bollino F. Characterization and biological properties of TiO_2/PCL hybrid layers prepared via sol-gel dip coating for surface modification of titanium implants. Journal of Non-Crystalline Solids. 2015;**415**:9-15

[14] Yu J, Zhao X, Zhao Q. Effect of surface structure on photocatalytic activity of TiO_2 thin films prepared by sol-gel method. Thin Solid Films. 2000;**379**(1-2):7-14

[15] Figueira R, Fontinha I, Silva C, Pereira E, Figueira RB, Fontinha IR, et al. Hybrid sol-gel coatings: Smart and green materials for corrosion mitigation. Coatings. 2016;**6**(1):12

[16] Aydemir S. Effects of withdrawal speed on the microstructural and optical properties of sol-gel grown ZnO:Al thin films. Vacuum. 2015;**120**:51-58

[17] Brinker CJ, Frye GC, Hurd AJ, Ashley CS. Fundamentals of

sol-gel dip coating. Thin Solid Films. 1991;**201**(1):97-108

[18] Nazeer AA, Madkour M. Potential use of smart coatings for corrosion protection of metals and alloys: A review. Journal of Molecular Liquids. 2018;**253**:11-22

[19] Hughes AE, Cole IS, Muster TH, Varley RJ. Designing green, self-healing coatings for metal protection. NPG Asia Materials. 2010;**2**(4):143-151

[20] Kodas T. Handbook of chemical vapor deposition (CVD), principles, technology, and applications. By Hugh O. Pierson, Noyes, Park Ridge, NJ, 1992. 436 pp., hardback, $ 68, ISBN 0-8155-1300-3, Advanced Materials. Vol. 5. USA: Noyes Publications; 1993. 401-402 p

[21] Wiiala UK, Penttinen IM, Korhonen AS, Aromaa J, Ristolainen E. Improved corrosion resistance of physical vapour deposition coated TiN and ZrN. Surface and Coatings Technology. 1990;**41**(2):191-204

[22] Liu Q, Chen D, Kang Z. One-step electrodeposition process to fabricate corrosion-resistant superhydrophobic surface on magnesium alloy. ACS Applied Materials & Interfaces. 2015;**7**(3):1859-1867

[23] Asri RIM, Harun WSW, Samykano M, Lah NAC, Ghani SAC, Tarlochan F, et al. Corrosion and surface modification on biocompatible metals: A review. Materials Science and Engineering: C. 2017;**77**:1261-1274

[24] Zheng S, Li J. Inorganic–organic sol gel hybrid coatings for corrosion protection of metals. Journal of Sol-Gel Science and Technology. 2010;**54**(2):174-187

[25] Figueira RB, Silva CJR, Pereira EV. Organic–inorganic hybrid sol–gel coatings for metal corrosion protection: A review of recent progress. Journal of Coatings Technology and Research. 2015;**12**(1):1-35

[26] Santana I, Pepe A, Schreiner W, Pellice S, Ceré S. Hybrid sol-gel coatings containing clay nanoparticles for corrosion protection of mild steel. Electrochimica Acta. 2016;**203**:396-403

[27] Dave BC, Hu X, Devaraj Y, Dhali SK. Sol-gel-derived corrosion-protection coatings. Journal of Sol-Gel Science and Technology. 2004;**32**(1-3):143-147

[28] Bera S, Rout TK, Udayabhanu G, Narayan R. Comparative study of corrosion protection of sol-gel coatings with different organic functionality on Al-2024 substrate. Progress in Organic Coatings. 2015;**88**:293-303

[29] Cambon J-B, Esteban J, Ansart F, Bonino J-P, Turq V, Santagneli SH, et al. Effect of cerium on structure modifications of a hybrid sol–gel coating, its mechanical properties and anti-corrosion behavior. Materials Research Bulletin. 2012;**47**(11):3170-3176

[30] Montemor MF. Functional and smart coatings for corrosion protection: A review of recent advances. Surface and Coatings Technology. 2014;**258**:17-37

[31] Vimalanandan A, Lv L-P, Tran TH, Landfester K, Crespy D, Rohwerder M. Redox-responsive self-healing for corrosion protection. Advanced Materials. 2013;**25**(48):6980-6984

[32] Sun D, An J, Wu G, Yang J. Double-layered reactive microcapsules with excellent thermal and non-polar solvent resistance for self-healing coatings. Journal of Materials Chemistry A. 2015;**3**(8):4435-4444

[33] Yabuki A. Self-healing coatings for corrosion inhibition of metals. Modern Applied Science. 2015;**9**(7):214

[34] Snihirova D, Lamaka SV, Taryba M, Salak AN, Kallip S, Zheludkevich ML, et al. Hydroxyapatite microparticles as feedback-active reservoirs of corrosion inhibitors. ACS Applied Materials & Interfaces. 2010;**2**(11):3011-3022

[35] Li GL, Zheng Z, Möhwald H, Shchukin DG. Silica/polymer double-walled hybrid nanotubes: Synthesis and application as stimuli-responsive nanocontainers in self-healing coatings. ACS Nano. 2013;**7**(3):2470-2478

[36] Manam NS, Harun WSW, Shri DNA, Ghani SAC, Kurniawan T, Ismail MH, et al. Study of corrosion in biocompatible metals for implants: A review. Journal of Alloys and Compounds. 2017;**701**:698-715

[37] Grainger DW. The Williams dictionary of biomaterials. Materials Today. 1999;**2**(3):29

[38] Ducheyne P. Comprehensive Biomaterials. Ankara, Turkey: Elsevier; 2011. 349-371 p

[39] Kurtz SM, Devine JN. PEEK biomaterials in trauma, orthopedic, and spinal implants. Biomaterials. 2007;**28**(32):4845-4869

[40] Li J, Hastings GW. Chapter 5. Oxide bioceramics: Inert ceramic materials in medicine and dentistry. In: Handbook of Biomaterial Properties. New York, NY: Springer New York; 2016. pp. 339-352

[41] Dorozhkin S. Calcium orthophosphate-containing biocomposites and hybrid biomaterials for biomedical applications. Journal of Functional Biomaterials. 2015;**6**(3):708-832

[42] Wen C. Surface Coating and Modification of Metallic Biomaterials. In: Wen C, editor. 1st ed. Sawston, Cambridge: Woodhead Publishing; 2015. 1-431 p

[43] Pilliar RM. Sol–gel surface modification of biomaterials. In: Wen C, editor. Surface Coating and Modification of Metallic Biomaterials. 1st ed. Sawston, Cambridge: Woodhead Publishing; 2015. p. 185-217

[44] Sidane D, Chicot D, Yala S, Ziani S, Khireddine H, Iost A, et al. Study of the mechanical behavior and corrosion resistance of hydroxyapatite sol–gel thin coatings on 316 L stainless steel pre-coated with titania film. Thin Solid Films. 2015;**593**:71-80

[45] Le Guéhennec L, Soueidan A, Layrolle P, Amouriq Y. Surface treatments of titanium dental implants for rapid osseointegration. Dental Materials. 2007;**23**(7):844-854

[46] Klokkevold PR, Nishimura RD, Adachi M, Caputo A. Osseointegration enhanced by chemical etching of the titanium surface. A torque removal study in the rabbit. Clinical Oral Implants Research. 1997;**8**(6):442-447

[47] Miklaszewski A, Jurczyk MU, Jurczyk K, Jurczyk M. Plasma surface modification of titanium by TiB precipitation for biomedical applications. Surface and Coatings Technology. 2011;**206**(2-3):330-337

[48] Liu X, Chu PK, Ding C. Surface modification of titanium, titanium alloys, and related materials for biomedical applications. Materials Science & Engineering R: Reports. 2004;**47**(3-4):49-121

[49] Parr GR. Tissue-integrated prostheses: Osseointegration in clinical dentistry. The Journal of Prosthetic Dentistry. 1985;**54**(4):611-612

[50] Holopainen J, Ritala M. Rapid production of bioactive hydroxyapatite fibers via electroblowing. Journal of the European Ceramic Society. 2016;**36**(13):3219-3224

[51] Hench LL. Bioceramics: From concept to clinic. Journal of the American Ceramic Society. 1991;**74**(7):1487-1510

[52] Bagambisa FB, Joos U, Schilli W. Mechanisms and structure of the bond between bone and hydroxyapatite ceramics. Journal of Biomedical Materials Research. 1993;**27**(8):1047-1055

[53] Mary IR, Sonia S, Viji S, Mangalaraj D, Viswanathan C, Ponpandian N. Novel multiform morphologies of hydroxyapatite: Synthesis and growth mechanism. Applied Surface Science. 2016;**361**:25-32

[54] Arends J, Christoffersen J, Christoffersen MR, Eckert H, Fowler BO, Heughebaert JC, et al. A calcium hydroxyapatite precipitated from an aqueous solution: An international multimethod analysis. Journal of Crystal Growth. 1987;**84**(3):515-532

[55] Osaka A, Miura Y, Takeuchi K, Asada M, Takahashi K. Calcium apatite prepared from calcium hydroxide and orthophosphoric acid. Journal of Materials Science. Materials in Medicine. 1991;**2**(1):51-55

[56] ŚAlósarczyk A, Stobierska E, Paszkiewicz Z, Gawlicki M. Calcium phosphate materials prepared from precipitates with various calcium:Phosphorus molar ratios. Journal of the American Ceramic Society. 2005;**79**(10):2539-2544

[57] Young RA, Holcomb DW. Variability of hydroxyapatite preparations. Calcified Tissue International. 1982;**34**(Suppl 2):S17-S32

[58] Royer A, Viguie JC, Heughebaert M, Heughebaert JC. Stoichiometry of hydroxyapatite: Influence on the flexural strength. Journal of Materials Science. Materials in Medicine. 1993;**4**(1):76-82

[59] Bloebaum RD, Dupont JA. Osteolysis from a press-fit hydroxyapatite-coated implant: A case study. Journal of Arthroplasty. 1993;**8**(2):195-202

[60] Jarcho M. Retrospective analysis of hydroxyapatite development for oral implant applications. Dental Clinics of North America. 1992;**36**(1):19-26

[61] Ben-Arfa BAE, Salvado IMM, Ferreira JMF, Pullar RC. Novel route for rapid sol-gel synthesis of hydroxyapatite, avoiding ageing and using fast drying with a 50-fold to 200-fold reduction in process time. Materials Science and Engineering: C. 2017;**70**:796-804

[62] Yoshimura M, Suda H, Okamoto K, Ioku K. Hydrothermal synthesis of biocompatible whiskers. Journal of Materials Science. 1994;**29**(13):3399-3402

[63] Chen W, Oh S, Ong AP, Oh N, Liu Y, Courtney HS, et al. Antibacterial and osteogenic properties of silver-containing hydroxyapatite coatings produced using a sol gel process. Journal of Biomedial Materials Research Part A. 2007;**82A**(4):899-906

[64] Albrektsson T, Sennerby L. State of the art in oral implants. Journal of Clinical Periodontology. 1991;**18**(6):474-481

[65] Johnson BW. HA-coated dental implants: Long-term consequences. Journal of the California Dental Association. 1992;**20**(6):33-41

[66] Truhlar RS, Morris HF, Ochi S. Implant surface coating and bone quality-related survival outcomes through 36 months post-placement of root-form endosseous dental implants. Annals of Periodontology. 2000;**5**(1):109-118

[67] de Bruijn JD, Bovell YP, van Blitterswijk CA. Structural arrangements at the interface between

plasma sprayed calcium phosphates and bone. Biomaterials. 1994;**15**(7):543-550

[68] Park E, Condrate RA. Graded coating of hydroxyapatite and titanium by atmospheric plasma spraying. Materials Letters. 1999;**40**(5):228-234

[69] Akao M, Aoki H, Kato K. Mechanical properties of sintered hydroxyapatite for prosthetic applications. Journal of Materials Science. 1981;**16**(3):809-812

[70] Tapsir Z, Saidin S. Synthesis and characterization of collagen–hydroxyapatite immobilized on polydopamine grafted stainless steel. Surface and Coatings Technology. 2016;**285**:11-16

[71] Mousa HM, Tiwari AP, Kim J, Adhikari SP, Park CH, Kim CS. A novel in situ deposition of hydroxyapatite nanoplates using anodization/hydrothermal process onto magnesium alloy surface towards third generation biomaterials. Materials Letters. 2016;**164**:144-147

[72] Fomin A, Dorozhkin S, Fomina M, Koshuro V, Rodionov I, Zakharevich A, et al. Composition, structure and mechanical properties of the titanium surface after induction heat treatment followed by modification with hydroxyapatite nanoparticles. Ceramics International. 2016;**42**(9):10838-10846

[73] Bandyopadhyay A, Sahasrabudhe H, Bose S. Laser surface modification of metallic biomaterials. In: Vilar R, editor. Laser Surface Modification of Biomaterials: Techniques and Applications. 1st ed. Sawston, Cambridge: Woodhead Publishing; 2016. p. 175-195

[74] Lin Z, Li S-J, Sun F, Ba D-C, Li X-C. Surface characteristics of a dental implant modified by low energy oxygen ion implantation. Surface and Coatings Technology. Elsevier; Sep 2018. DOI: 10.1016/j.surfcoat.2018.09.003

[75] Lee WD, Gawri R, Pilliar RM, Stanford WL, Kandel RA. Sol gel-derived hydroxyapatite films over porous calcium polyphosphate substrates for improved tissue engineering of osteochondral-like constructs. Acta Biomaterialia. 2017;**62**:352-361

[76] Gupta R, Kumar A. Bioactive materials for biomedical applications using sol-gel technology. Biomedical Materials. 2008;**3**(3):034005

[77] Asri RIM, Harun WSW, Hassan MA, Ghani SAC, Buyong Z. A review of hydroxyapatite-based coating techniques: Sol–gel and electrochemical depositions on biocompatible metals. Journal of the Mechanical Behavior of Biomedical Materials. 2016;**57**:95-108

[78] El Hadad A, Peón E, García-Galván F, Barranco V, Parra J, Jiménez-Morales A, et al. Biocompatibility and corrosion protection behaviour of hydroxyapatite sol-gel-derived coatings on Ti6Al4V alloy. Materials (Basel). 2017;**10**(2):94

[79] Jokinen M, Pätsi M, Rahiala H, Peltola T, Ritala M, Rosenholm JB. Influence of sol and surface properties on in vitro bioactivity of sol-gel-derived TiO_2 and TiO_2-SiO_2 films deposited by dip-coating method. Journal of Biomedical Materials Research. 1998;**42**(2):295-302

[80] Albert M, Clifford A, Zhitomirsky I, Rubel O. Adsorption of maleic acid monomer on the surface of hydroxyapatite and TiO_2 : A pathway toward biomaterial composites. ACS Applied Materials & Interfaces. 2018;**10**(29):24382-24391

[81] Hay JN, Raval HM. Synthesis of organic-inorganic hybrids via the non-hydrolytic sol-gel process. Chemistry of Materials. 2001;**13**:3396-3403

[82] Takahashi Y, Matsuoka Y. Dip-coating of TiO_2 films using a sol derived from Ti(O-i-Pr)4-diethanolamine-H_2O-i-PrOH system. Journal of Materials Science. 1988;**23**(6):2259-2266

[83] Trentler TJ, Denler TE, Bertone JF, Agrawal A, Colvin VL. Synthesis of TiO_2 nanocrystals by nonhydrolytic solution-based reactions. Journal of the American Chemical Society. 1999;**121**:1613-1614

[84] Shen GX, Chen YC, Lin CJ. Corrosion protection of 316 L stainless steel by a TiO_2 nanoparticle coating prepared by sol-gel method. Thin Solid Films. 2005;**489**(1-2):130-136

[85] Balamurugan A, Kannan S, Rajeswari S. Evaluation of TiO_2 coatings obtained using the sol–gel technique on surgical grade type 316L stainless steel in simulated body fluid. Materials Letters. 2005;**59**(24-25):3138-3143

[86] Vasconcelos DCL, Nunes EHM, Sabioni ACS, da Costa JCD, Vasconcelos WL. Structural characterization and corrosion behavior of stainless steel coated with sol-gel titania. Journal of Materials Engineering and Performance. 2012;**21**(3):411-417

[87] Burnat B, Dercz G, Blaszczyk T. Structural analysis and corrosion studies on an ISO 5832-9 biomedical alloy with TiO_2 sol–gel layers. Journal of Materials Science. Materials in Medicine. 2014;**25**(3):623-634

[88] Burnat B, Blaszczyk T, Leniart A. Effects of serum proteins on corrosion behavior of ISO 5832-9 alloy modified by titania coatings. Journal of Solid State Electrochemistry. 2014;**18**(11):3111-3119

[89] Peres-Durand S, Rouviere J, Guizard C. Sol-gel processing of titania using reverse micellar systems as reaction media. Colloids and Surfaces A: Physicochemical and Engineering Aspects. 1995;**98**(3): 251-270

[90] Shimizu T, Fujibayashi S, Yamaguchi S, Yamamoto K, Otsuki B, Takemoto M, et al. Bioactivity of sol–gel-derived TiO_2 coating on polyetheretherketone: In vitro and in vivo studies. Acta Biomaterialia. 2016;**35**:305-317

[91] Sakka S. Preparation and properties of sol-gel coating films. Journal of Sol-Gel Science and Technology. 1994; **2**(1-3):451-455

[92] Liu J-X, Yang D-Z, Shi F, Cai Y-J. Sol–gel deposited TiO_2 film on NiTi surgical alloy for biocompatibility improvement. Thin Solid Films. 2003;**429**(1-2):225-230

[93] Hu J, Zhang C, Cui B, Bai K, Guan S, Wang L, et al. In vitro degradation of AZ31 magnesium alloy coated with nano TiO_2 film by sol–gel method. Applied Surface Science. 2011;**257**(21): 8772-8777

[94] Burnat B, Robak J, Leniart A, Piwoński I, Batory D. The effect of concentration and source of calcium ions on anticorrosion properties of Ca-doped TiO_2 bioactive sol-gel coatings. Ceramics International. 2017;**43**(16):13735-13742

[95] Hench LL. The story of Bioglass®. Journal of Materials Science: Materials in Medicine. 22 Nov 2006;**17**(11):967-978

[96] Barrioni BR, Aline A, Oliveira RD. The Evolution, Control, and Effects of the Compositions of Bioactive Glasses on Their Properties and Applications. In: Marchi J, editor. Biocompatible Glasses From Bone Regeneration to Cancer Treatment. 1st ed. Switzerland: Springer International Publishing; 2016. p. 85-117

[97] Zhong J, Greenspan DC. Processing and properties of sol-gel bioactive

glasses. Journal of Biomedical Materials Research. 2000;**53**(6):694-701

[98] Catauro M, Papale F, Sapio L, Naviglio S. Biological influence of Ca/P ratio on calcium phosphate coatings by sol-gel processing. Materials Science & Engineering. C, Materials for Biological Applications. 2016;**65**:188-193

[99] Andersson H, Kangasniemi I. Calcium phosphate formation at the surface of bioactive glass in vitro. Journal of Biomedical Materials Research. 1991;**25**(8):1019-1030

[100] Hench LL, Wilson J. Surface-active biomaterials. Science. 1984;**226**(4675):630-636

[101] Wallace KE, Hill RG, Pembroke JT, Brown CJ, Hatton P. Influence of sodium oxide content on bioactive glass properties. Journal of Materials Science. Materials in Medicine. 1999;**10**(12):697-701

[102] Ducheyne P, Qiu Q. Bioactive ceramics: The effect of surface reactivity on bone formation and bone cell function. Biomaterials. 1999;**20**:2287-2303

[103] Krishnan V, Lakshmi T. Bioglass: A novel biocompatible innovation. Journal of Advanced Pharmaceutical Technology & Research. 2013;**4**(2):78-83

[104] Shirtliff VJ, Hench LL. Bioactive materials for tissue engineering, regeneration and repair. Journal of Materials Science. 2003;**38**(23):4697-4707

[105] Sepulveda P, Jones JR, Hench LL. Characterization of melt-derived 45S5 and sol-gel-derived 58S bioactive glasses. Journal of Biomedical Materials Research. 2001;**58**(6):734-740

[106] Shen S, Cai S, Xu G, Zhao H, Niu S, Zhang R. Influence of heat treatment on bond strength and corrosion resistance of sol–gel derived bioglass–ceramic coatings on magnesium alloy. Journal of the Mechanical Behavior of Biomedical Materials. 2015;**45**:166-174

[107] Meyer N, Rivera L, Ellis T, Qi J, Ryan M, Boccaccini A, et al. Bioactive and antibacterial coatings based on zein/bioactive glass composites by electrophoretic deposition. Coatings. 2018;**8**(1):27

[108] Xue B, Guo L, Chen X, Fan Y, Ren X, Li B, et al. Electrophoretic deposition and laser cladding of bioglass coating on Ti. Journal of Alloys and Compounds. 2017;**710**:663-669

[109] Rocha RC, Galdino AG de S, da Silva SN, Machado MLP. Surface, microstructural, and adhesion strength investigations of a bioactive hydroxyapatite-titanium oxide ceramic coating applied to Ti-6Al-4V alloys by plasma thermal spraying. Materials Research. May 28 2018;**21**(4):1-8

[110] Juhasz JA, Best SM. Bioactive ceramics: Processing, structures and properties. Journal of Materials Science. 2012;**47**(2):610-624

[111] Sola A, Bellucci D, Cannillo V, Cattini A. Bioactive glass coatings: A review. Surface Engineering. 2011;**27**(8):560-572

[112] Fathi MH, Doostmohammadi A. Bioactive glass nanopowder and bioglass coating for biocompatibility improvement of metallic implant. Journal of Materials Processing Technology. 2009;**209**(3):1385-1391

[113] Wang D, Bierwagen GP. Sol–gel coatings on metals for corrosion protection. Progress in Organic Coatings. 2009;**64**(4):327-338

[114] Hornberger H, Virtanen S, Boccaccini AR. Biomedical coatings on magnesium alloys—A review. Acta Biomaterialia. 2012;**8**(7):2442-2455

[115] Sato M, Slamovich EB, Webster TJ. Enhanced osteoblast adhesion on hydrothermally treated hydroxyapatite/titania/poly(lactide-co-glycolide) sol-gel titanium coatings. Biomaterials. 2005;**26**(12):1349-1357

[116] Catauro M, Bollino F, Papale F, Ferrara C, Mustarelli P. Silica–polyethylene glycol hybrids synthesized by sol–gel: Biocompatibility improvement of titanium implants by coating. Materials Science and Engineering: C. 2015;**55**:118-125

[117] Mehdipour M, Afshar A. A study of the electrophoretic deposition of bioactive glass–chitosan composite coating. Ceramics International. 2012;**38**(1):471-476

[118] Juan-Díaz MJJ, Martínez-Ibáñez M, Lara-Sáez I, da Silva S, Izquierdo R, Gurruchaga M, et al. Development of hybrid sol–gel coatings for the improvement of metallic biomaterials performance. Progress in Organic Coatings. 2015;**96**:42-51

[119] Huang Y, Zhang X, Zhang H, Qiao H, Zhang X, Jia T, et al. Fabrication of silver- and strontium-doped hydroxyapatite/TiO_2 nanotube bilayer coatings for enhancing bactericidal effect and osteoinductivity. Ceramics International. 2017;**43**(1):992-1007

[120] Córdoba LC, Hélary C, Montemor F, Coradin T. Bi-layered silane-TiO_2/collagen coating to control biodegradation and biointegration of Mg alloys. Materials Science and Engineering: C. 2019;**94**:126-138

[121] Jun S-H, Lee E-J, Yook S-W, Kim H-E, Kim H-W, Koh Y-H. A bioactive coating of a silica xerogel/chitosan hybrid on titanium by a room temperature sol–gel process. Acta Biomaterialia. 2010;**6**(1):302-307

[122] Catauro M, Bollino F, Papale F, Mozetic P, Rainer A, Trombetta M. Biological response of human mesenchymal stromal cells to titanium grade 4 implants coated with PCL/ZrO_2 hybrid materials synthesized by sol–gel route: In vitro evaluation. Materials Science and Engineering: C. 2014;**45**:395-401

[123] Catauro M, Bollino F, Papale F, Gallicchio M, Pacifico S. Influence of the polymer amount on bioactivity and biocompatibility of SiO_2/PEG hybrid materials synthesized by sol-gel technique. Materials Science and Engineering: C. 2015;**48**:548-555

[124] Catauro M, Bollino F, Papale F. Preparation, characterization, and biological properties of organic-inorganic nanocomposite coatings on titanium substrates prepared by sol-gel. Journal of Biomedial Materials Research Part A. 2014;**102**(2):392-399

Chapter 4

Synthesis, Characterization, and Adsorption Properties of Nanoporous Materials

Rolando Roque-Malherbe and Carlos de las Pozas del Rio

Abstract

During the last years, the authors have synthesized, characterized, and studied the adsorption properties of nitroprussides, Prussian blue analogues, akaganeites, MeAPOs, metal–organic frameworks, and extremely high specific surface amorphous silica, which allowed the storage of about 11 wt.% of hydrogen in the form of ammonia. In this sense, using the solid-state reaction method, sol–gel methodologies, together with aluminosilicate, high silica and non-aluminosilicate zeolite synthesis methods, were described, moreover was explained how to prepare active carbons along with the synthesis of Prussian blue analogues (PBAs) and nitroprussides (NPs). In addition, the characterization of the materials of interest applying X-ray diffraction, thermogravimetric analysis, DRIFTS, and room-temperature Mossbauer spectrometry was discussed. Besides, the concepts that define physical adsorption and examples of adsorption data, which were tested with the help of the Dubinin, osmotic adsorption and Langmuir-type isotherms, were defined. Later, the methodology was described for the measurement of adsorption data with the help of the volumetric method. Moreover, a description of the thermodynamics of adsorption, along with the methodology for the calculation of calorimetric data with the help of heat flow calorimeters together with the measurement of differential heats of adsorption data was developed. Finally, the different interaction forces that make possible adsorption were discussed.

Keywords: physical adsorption, isotherms, Dubinin and osmotic tests, differential heats of adsorption, nitroprussides, Prussian blue analogues, silica and natural zeolites

1. Introduction

The concept of "adsorption" was recommended by Kayser in 1881 to explain the growth in concentration of gas molecules on the adjacent surface of a solid adsorbent, an effect previously noted by Fontana and Scheele in 1777 [1]. In particular, majority of adsorbents useful for industrial applications have pores with hole dimensions in the nanometer region; in this pore-size zone, adsorption is on the one hand a significant methodology for the characterization of porous materials; to be exact, gas adsorption offers evidence concerning the mesopore area, volume and size of the pores together with the energetics of adsorption [2]. Also, gas adsorption is a vital unitary operation for sustainable energy and pollution abatement

IntechOpen

applications of nanoporous materials along with the industrial uses, where materials like metal organic framework molecular sieves, metal organic frameworks, Prussian blue analogues, mesoporous molecular sieves, pillared clays, silica, alumina, active carbons, titanium dioxides, magnesium oxides, carbon nanotubes together with zeolites and related materials are the most widely studied and applied adsorbents in Science and Technology [3].

In addition to their adsorption properties, nanoporous materials are a group of advanced materials with other excellent properties and applications in many fields, for example, waste water treatment [4–6], ionic conduction [7], ionic exchange [8], gas separation [9], membranes [10], catalysts [11], catalyst supports [12] and detergency [13].

The classification of the different pore widths of porous adsorbents was carried out by the International Union of Pure and Applied Chemistry (IUPAC); in this classification, adsorbent materials are categorized as those with pore diameters greater than 50 nm, named macroporous adsorbents, while mesoporous materials show pore diameters between 2 and 50 nm that are termed microporous; those adsorbents showing pore diameters between 0.3 and 2 nm, where, the pore width, *Dp,* is defined as equal to the diameter in the case of cylindrical shaped pores, and as the distance between opposite walls in the case of slit-shaped pores [14].

In addition, the clean surface of an adsorbent is characterized by the fact that the atoms that produce the surface have non-saturated bonds, fact producing an adsorption field above this surface; been the adsorption field the cause of the formation of a stockpile of molecules close to the adsorbent surface; this effect, i.e., adsorption, is a universal propensity of surface systems, given that throughout this process a reduction of the surface tension is experienced by the solid; been, adsorption the term applied to describe this process; whereas for the opposite, the term desorption is used [1].

Now it is necessary to state that the occupation of the adsorption space by adsorbed molecules in complex porous systems occurs roughly in the following form: initially, micropore filling takes place, with the adsorption process being controlled almost totally by the interaction of the adsorbed molecules with the pore walls; thereafter, at higher pressures, the external surface is covered, which is a monolayer and a multilayer adsorption on the walls of mesopores, and open macropores take place. Finally, capillary condensation occurs in the mesopores [2]. On the other hand, dynamic adsorption is a mass transfer between a mobile, solid, or liquid phase, and the adsorption bed packed in a reactor, been necessary to carry out the dynamic adsorption, a reactor, where the adsorption process will occur in the reactor adsorbent packed bed, been, the adsorbents normally used for these applications are active carbons, zeolites and related materials, silica, mesoporous molecular sieves, alumina, titanium dioxide, magnesium oxide, clays, and pillared clays [3].

2. Synthesis

2.1 Solid-state reaction method

Inorganic compounds are mixed, following stoichiometric proportions; after that, the powders are thoroughly milled and then finally thermally treated to get the required product, been, perovskites synthesized using this method [15, 16].

2.2 Sol-gel methodologies

These processes can be based on the hydrolysis condensation of metal alkoxides. In the important case of amorphous silica synthesis, silicic acid is first produced by the hydrolysis of a silicon alkoxide, formally a silicic acid ether; then, the global reaction continues as a condensation polymerization to form high molecular weight polysilicates, which connect together to produce a gel [17].

2.3 Aluminosilicate zeolites

These materials are normally synthesized in hydrothermal conditions using solutions that are composed of sodium hydroxide, sodium silicate and sodium aluminate, where three steps are observed during the synthesis process, i.e., induction, nucleation, and crystallization, which determine the specific zeolite produced by the applied reactants, along with the parameters used, such as temperature, pH, and time [18, 19].

2.4 High silica and non-aluminosilicate zeolites

In this case, the aluminosilicate synthesis is complemented with the addition of structure-directing agents (SDA), such as quaternary ammonium cations, linear or cyclic ethers, and coordination compounds. been, the first high-silica zeolites, that is, ETA, EU, NU, and the ZSM series, patented in the late 1960s, or early 1980s [20]; moreover, high silica zeolite can be as well synthesized using ethanol and seeds of the desired phase [21]; meanwhile, Microporous aluminophosphate molecular sieves, that is, non-aluminosilicate zeolites; such as: SAPO molecular sieves were obtained by incorporation of Si in the AlPO framework, while MeAPO molecular sieves are obtained by the inclusion of Ga, Be, Va, Co, Fe, Mg, Mn, Zn in the AlPO framework obtaining between others: gallophosphates, zincophosphates, beryllophosphates, vanadophosphates and ferrophosphates [22].

Active Carbon prepared with wood, lignite, peat, coconut, eucalyptus lignin, and apricot, cherry and olive stones under physical (a) and chemical activation (b) methods; i.e. [23–25]:

a. carbonization at 800–1000°C under inert gas, and activation under oxidizing agents, as a rule, carbon dioxide or water vapor.

b. chemical activation requires a treatment with: sulfuric acid, phosphoric acid, zinc chloride, potassium hydroxide at 400–1000°C, followed by the elimination of the dehydrating agent by meticulous washing.

Prussian blue analogues (PBAs) [26, 27] and nitroprussides (NPs) [28] are produced by mixing solutions of $K_3[Fe(CN)_6]$, for the synthesis of Prussian blue analogues, and $Na_2[Fe(CN)_5NO]$ for the synthesis of nitroprussides with a nitrate of the corresponding metal to get the corresponding PBAs and NPs.

3. Characterization

The X-ray Diffraction (XRD) methodology was used to determine the crystalline phases present in the natural and synthesized materials, together with the investigation of the nucleation and growth process during the synthesis of different materials to be used as adsorbents, ion exchangers and catalysts [29]. The XRD tests were carried out, at room temperature using a stage similar to an Anton Paar HTK-1200 N with an equipment similar to a Bruker D8 Advance system in a Bragg-Brentano vertical goniometer configuration, the angular measurements being made applying steps of 0.01° from 5 to 80°, using a Cu anode tube together with a Ni filter placed, prior to the detector, to eliminate Cu K_α radiation the one-dimensional detector was employed [26].

3.1 Thermo-gravimetric analysis (TGA)

The analytical test was carried out with an equipment similar to a TA, Q-500 instrument; the samples were placed onto a ceramic sample holder suspended from an analytical balance, suspended from an analytical balance; then the temperature was linearly scanned, from 25 to 300 C, at a heating rate of 5 C/min under a pure N_2 flow of 100 mL/min. TGA testing process was carried out with a TA, Q-500 instrument, and the samples were placed onto a ceramic sample holder [27].

DRIFTS were gathered using equipment similar to a Thermo Scientific Nicolet iS10 FTIR spectrometer. The data of the hydrated and dehydrated samples were collected at a resolution of 4 cm^{-1} employing 100 scans per sample; a background, with pure KBr, was in all cases collected before the collection of the spectra. Both the hydrated and dehydrated samples were gathered at room temperature while flowing N_2 (Praxair, 99.99%) at a rate of 50 cc/min; while, the dehydration of the studied samples was performed at 100°C for 2 h in N_2 flow of 50 mL/min while the sample under test was included in the finger-sample-holder of the IR high temperature cell [28].

Room-temperature Mossbauer Spectrometry measurements were made with a system similar to a SEECo supplied spectrometer operating at constant acceleration mode with a 50 mCi ^{57}Co γ-ray source in a Rh matrix made by Rietverc GmbH. The 1024-point raw data were folded and analyzed using WMOSS, a public domain Mossbauer spectral analysis program; the calibration was made with reference to α-Fe metal [30].

4. Definition and terms

4.1 Adsorption isotherm

It is the most important measurement made to characterize an adsorbent; it is defined as the relationship between the amount adsorbed, n_a, and the equilibrium pressure, P, at constant temperature, T [1, 2] (**Figure 1**)

$$n_a = f(P)_T \tag{1}$$

4.2 Physical and chemical adsorption

The gas adsorption process is normally considered a physical process, named physical adsorption, since the molecular forces involved in this process are usually of the van der Waals type. Meanwhile Chemical adsorption of gases in solid surfaces takes place in the case where during the adsorption process a reaction with exchange of electrons between the solid surface and the gas molecules with the formation of chemical bonds takes place, been, the adsorbent the solid phase and the adsorbate the gas phase [14].

4.3 Mobile and immobile adsorptions

Gas and vapor adsorption in solid adsorbents can be also classified as mobile adsorption, where the first case occurs when the adsorbed molecule is performed as

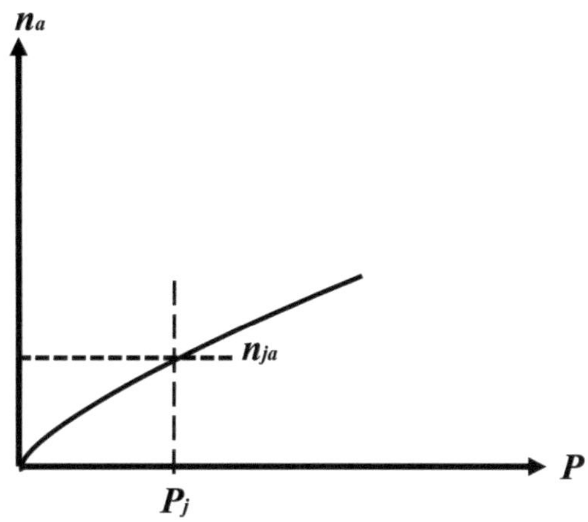

Figure 1.
Graphic representation of the adsorption isotherm.

a gas molecule is free to move within the adsorption space, or the instance of immobile adsorption, taking place in the case where the adsorbed molecule vibrates around an adsorption site [2, 3].

4.4 Monolayers and multilayers

For open surfaces, adsorption consists of a layer-by-layer loading process, where the first layer is filled as in the case when $\theta = \frac{n_a}{N_m} = 1$, where θ is the surface recovery and N_m is the monolayer capacity; as a result, it is understood that we have monolayer adsorption when $\theta < 1$ and multiplayer adsorption when $\theta > 1$ [31].

4.5 Porous adsorbents

They are characterized by their specific surface area, denoted by S, and measured in (m^2/g), in which the surface area is the outer surface, concretely, the area external to the micropores. If the adsorbents do not have micropores, the surface area and the outer surface area match; furthermore, the micropore volume is represented by W_{MP}, measured in (cm^3/g), whereas the total pore volume labeled W is the sum of the micropores and mesopore volumes [32] of the adsorbent, also measured in (cm^3/g). Lastly, we have the pore size distribution (PSD), i.e., a plot of $\frac{\Delta v_P}{\Delta D_p}$ versus *Dp*, where Vp is the pore volume accumulated up to the pore width *Dp* measured in (cc-STP/g Å) [26, 27]. Now it is necessary to state that cc-STP is a unit denoting the amount adsorbed, measured in cubic centimeters at standard temperature and pressure (STP), that is, 273.15 K and 760 Torr, that is, 1.01325×10^5 Pa.

4.6 Magnitude of adsorption

The interfacial layer is the non-homogeneous section of an adsorption system, i.e., between two neighboring bulk phases (**Figure 2**), where the properties typifying this region are radically dissimilar from, but associated to, the properties of the bulk phases; in which, to deal with this system is assumed that in the ideal reference system, the concentration must be constant up to the Gibbs dividing surface (GDS) (see **Figure 2**); been, in an actual system [32]

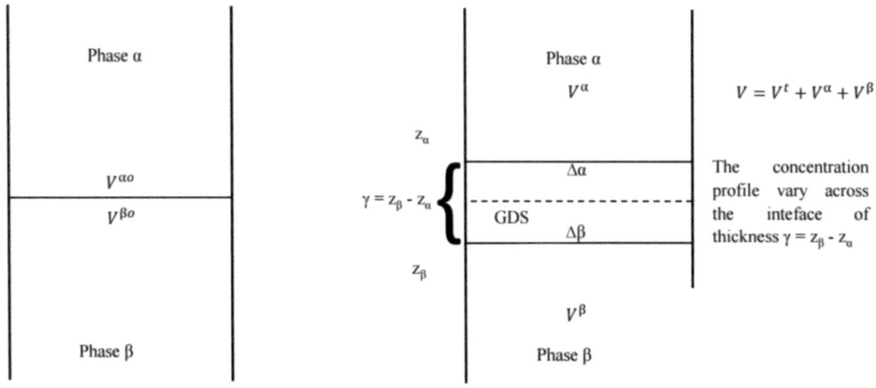

Figure 2.
Gibbs dividing surface.

$$n^a = \int_0^{V^a} cdV = \int_0^t cdz \tag{2}$$

$$V^\propto = At$$

Then the total amount of gas molecules in the system measured in (mol/g) is

$$n = n^a + V^g c_g^g$$

$$n^a = n - V^g c_g^g$$

where

$V^\propto$ is the volume of the adsorption space, V^g is the volume of the gas phase, c_g^g is the concentration of the gas phase, c_s^g is the concentration of the gas in the solid phase, A is the adsorbent surface area, t is the thickness of the surface layer, and n is the total amount of gas molecules in the system.

Finally the isotherm is calculated as follows:

$$n_a = \frac{n^a}{m_s} = F(P)_T$$

where m_s is the degassed adsorbent mass in grams. Then if the ideal gas equation is applied,

$$\mathrm{PV = nRT} \tag{3}$$

Been, n and T, kept constants, then [29]:

$$\mathrm{P_iV_i = P_jV_j = nRT = const} \tag{4}$$

where P is the pressure in atmosphere, V is the volume in milliliters; $1\ \mathrm{ml} = 1\ \mathrm{cm}^3 = 10^3\mathrm{mm}^3 = 10^{-3}\mathrm{dm}^3 = 10^{-3}\mathrm{l}$, n is the number of gas moles, T is the temperature in kelvin (K), and R is the gas constant.

Thereafter, using Eq. (4), based in the scheme of the volumetric adsorption equipment shown in **Figure 11**, we get:

$$\mathrm{P}_{at}\mathrm{V_m} = \mathrm{P_i(V_m + V_d + V_x)} \tag{5}$$

$$\mathrm{P}_{at} = 2l_{at}$$

In **Figure 11**, a schematic representation of volumetric gas adsorption measurement equipment is shown, where using the scheme, it is possible to show that:

$$V_x = \pi r^2 \left(X_i^{;} + l_0\right) \tag{6}$$

$$\left(X_i^{;} + l_0\right) = X_i$$

Initially V_m is filled with air and then the valves are opened allowing air up to equilibrium with the atmospheric pressure of the room, which can be measured with a barometer; now given that V_m was calibrated, filling it with water, then X_i allow us to calculate:

$$P_i[\text{mmHg}] = 2X_i[\text{mmHg}]$$

$$P_{at} = 2l_{at}[\text{mmHg}]$$

where P_{at} is the atmospheric pressure and P_1 is the measurement of pressure after expansion.

Substituting Eq. (6) on (5), the dose volume is calculated as:

$$V_d = \frac{l_{at}}{X_i} V_m - \left(V_m + \pi r^2 X_i\right)$$

In this sense, using the proportions rule, we get:

$$\frac{2X_i[\text{mmHg}]x1[\text{Atm}]}{760[\text{mmHg}]} = P_i[\text{Atm}]$$

Consequently applying the ideal gas law:

$$P_i[\text{Atm}]\left(2\pi r^2 X_i\left[\text{mm}^3\right]x10^3 + V_d[\text{ml}]\right) = n_i[\text{mmol}]R\left[\frac{\text{atm.ml}}{\text{mmol.K}}\right]T[K]$$

where

$$\frac{P_i[\text{Atm}]\left(2\pi r^2 X_i[\text{mm}^3]x10^3 + V_d[\text{ml}]\right)}{R\left[\frac{\text{atm.ml}}{\text{mmol.K}}\right]T[K]} = n_i[\text{mmol}]$$

Number of moles in the dose volume and the volume provided by the displacement of the manometer.

5. Experimental determination of V_a and V'_a and the numbers of mole gas adsorbed n_{aj}

When the valve of the sample holder is opened, the gas is expanded; hence:

$$\frac{P_i[\text{Atm}]\left(2\pi r^2 X_i[\text{mm}^3]x10^3 + V_d[\text{ml}]\right)}{R\left[\frac{\text{atm.ml}}{\text{mmol.K}}\right]T[K]} = \frac{P_j[\text{Atm}]\left(2\pi r^2 X_j[\text{mm}^3]x10^3 + V_d[\text{ml}] + V_a\right)}{R\left[\frac{\text{atm.ml}}{\text{mmol.K}}\right]T[K]} = n_i$$

where:

$$\frac{P_j}{P_i}[\text{Atm}]\left(2\pi r^2 X_j\left[\text{mm}^3\right]x10^3 + V_d[\text{ml}] + V_a\right) - [\text{Atm}]\left(2\pi r^2 X_i\left[\text{mm}^3\right]x10^3 + V_d[\text{ml}]\right) = V_a$$

This allows the calculation of Va, the so-called dead volume. Nevertheless the really expanded volume should consider the volume occupied by the adsorbent sample, i.e.:

$$V_{muestra}\left[cm^3\right] = \frac{m[g]}{\rho\left[\frac{g}{cm^3}\right]}$$

And ρ is the apparent density of the tested adsorbent material. Consequently:

$$V_a - V_{muestra} = V'_a$$

According ideal gas low:

$$\frac{P_i[Atm]\left(2\pi r^2 X_i[mm^3]x10^3 + V_d[ml]\right) + P_{j-1}V'_a}{R\left[\frac{atm.ml}{mmol.K}\right]T[K]} - \frac{P_j[Atm]\left(2\pi r^2 X_j[mm^3]x10^3 + V_d[ml] + V'_a\right)}{R\left[\frac{atm.ml}{mmol.K}\right]T[K]} = n_i - n_j$$

where:

$$n_i - n_j = n_{aj}$$

which is the amount adsorbed in the tested sample, which plotted against the equilibrium pressure P_j provides the adsorption isotherm.

6. The Dubinin adsorption isotherm equation

The Dubinin adsorption isotherm equation can be inferred by the application of the Dubinin's Theory of Volume Filling, i.e., (where volume filling is a process which takes place by the filling of the adsorption space rather than the surface coverage), together with, the Polanyi's adsorption potential [32]. In this regard, in 1914, Polanyi created perhaps the first convincing physical adsorption model. Mijail. M. Dubinin, a former pupil of Polanyi applied this model that essentially entailed a link between the adsorption space volume, V_i, and the adsorption energy field, ε_i, (see **Figure 3**): i.e.,$\varepsilon_i = F(V_i)$, termed by Polanyi as the characteristic function, a temperature independent function. For the deduction of the Dubinin isotherm equation, the next consideration was that [33]:

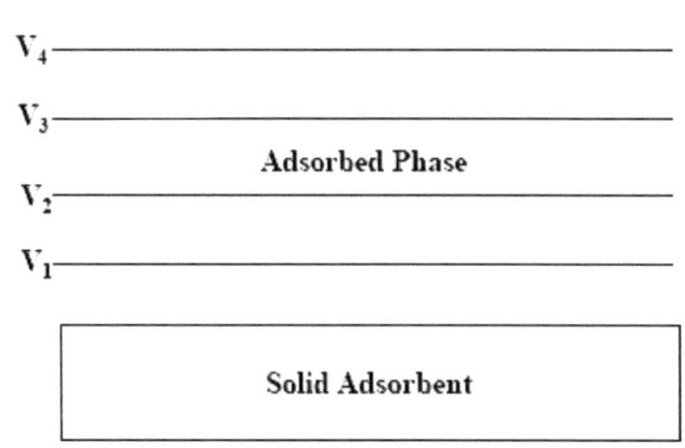

Figure 3.
Polanyi adsorption model.

$$\mu_g = \mu_L + \varepsilon_i = \mu_a \tag{7}$$

In which, μ_g is the chemical potential of the gas phase adsorbate, μ_L is the chemical potential of the pure liquid adsorbate, μ_a is the adsorbed phase chemical potential, while ε_i is the potential energy of the adsorption field. Consequently, applying of Eq. (7), it is possible to demonstrate that:

$$\varepsilon_i = RT \ln\left(\frac{P_0}{P_i}\right) \tag{8}$$

where P_o is the vapor pressure of the adsorptive at the temperature, T is the adsorption experiment, while P_i is the equilibrium adsorption pressure (ε could be also designed as the differential work of adsorption). Now, following the so-called Gurvich rule, it is possible to obtain the subsequent relation $V_i = V^L n_a$ between the volume of the adsorption space, V_i, and the amount adsorbed, where V^L is the molar volume of the liquid phase that conforms to the adsorbed phase. Combining Eq. (8) with the characteristic function, the relation between the volume of the adsorption space, and the amount adsorbed, we will get [32]:

$$F(V_i) = f(n_a) = \varepsilon_i = RT \ln\left(\frac{P_0}{P_i}\right) \tag{9}$$

Now, applying the Weibull distribution function, the relation between the amount adsorbed, n_a, and the differential work of adsorption, ε, is defined by the following relation [33]:

$$n_a = N_a \exp\left(-\frac{\varepsilon}{E}\right)^n \tag{10}$$

where E is a parameter termed the characteristic energy of adsorption; meanwhile, N_a is the maximum amount adsorbed in the volume of the micropore, n ($1 < n < 5$) being an empirical parameter. Now combining Eqs. (9) and (10) is feasible to construe the Dubinin adsorption isotherm equation [32] as follows:

$$n_a = N_a \exp\left(-\frac{RT}{E} \ln\left[\frac{P_0}{P}\right]\right)^n \tag{11}$$

It is possible, as well, to express the Dubinin adsorption isotherm equation in linear form:

$$\ln(n_a) = \ln(N_a) - \left(\frac{RT}{E}\right)^n \ln\left(\frac{P_0}{P}\right)^n \tag{12}$$

which is a very powerful tool for the description of the experimental data of adsorption in microporous material.

In **Figure 4** is shown the Dubinin plot of N_2 adsorption data at 77 K in the pressure range: $0.001 < P/Po < 0.03$, in a high silica commercial H-Y zeolite, precisely, the acid Y zeolite labeled CBV-720, manufactured by PQ corporation; where, adsorption data was gathered in an Autosorb-1 automatic volumetric gas adsorption system [26]; been evidently the experimental data correctly fitted by Eq. (12)

Recapitulating, the concrete form to made the linear Dubinin plot as represented in **Figure 4** was as follows:

$$y = \ln(n_a) = \ln(N_a) - \left(\frac{RT}{E}\right)^n \ln\left(\frac{P_0}{P}\right)^n = b - mx$$

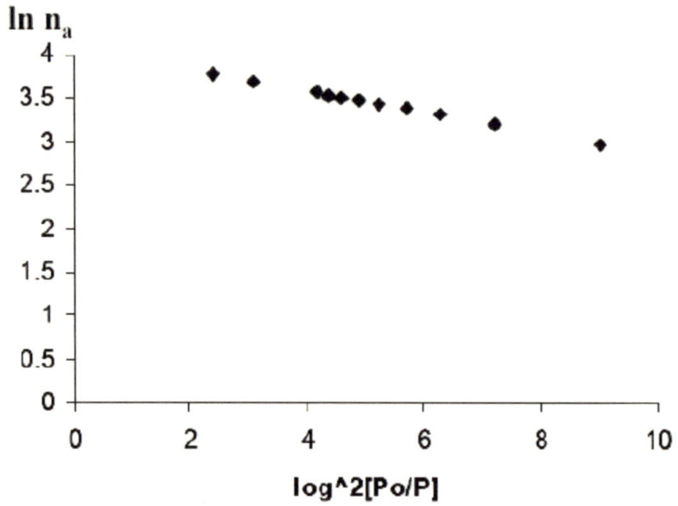

Figure 4.
Dubinin plot sample CBV-720 N_2 at 77 K [32].

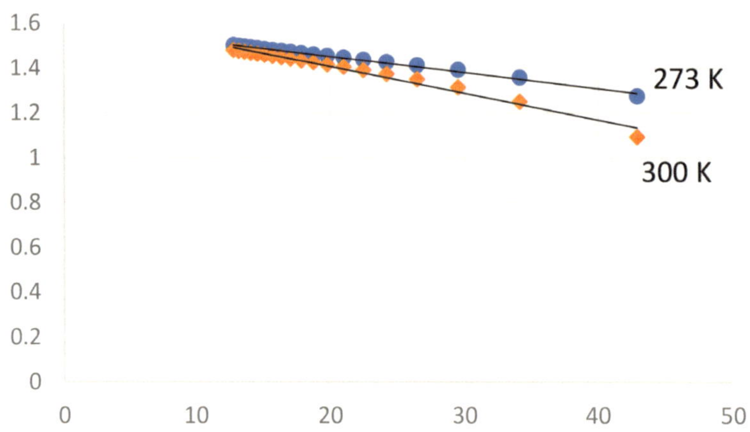

Figure 5.
Dubinin plot of carbon dioxide adsorption on nickel nitroprusside [34].

where $y = \ln(n_a)$, $b = \ln(N_a)$, $m = \left(\frac{RT}{E}\right)^n$, and $x = \ln\left(\frac{P_0}{P}\right)^n$

Nevertheless, this fitting process can be as well made using a non-linear regression method; in which, the fitting process is carried out with a program based on a least square procedure, allowing the calculation of the best fitting parameters of the Eq. (12), that is, N_a, E, been n, taken as a constant, for example, $n = 2$; additionally the program compute the regression coefficient, along with the standard errors.

As another example, in **Figure 4**, the use of the Dubinin equation in the measurement of the micropore volume, of a nickel nitroprusside (Ni-NP), is reported [34] (**Figure 5**).

7. Osmotic and Langmuir-type adsorption isotherms

Within the frame of the osmotic theory of adsorption, the adsorption phenomenon in a microporous adsorbent, for example a zeolite, is considered as the "osmotic" equilibrium between two solutions, i.e., vacancy plus molecules of different concentrations and the molecules in the gas phase. The solutions are formed in the micropores, and the gas phase, where the solvents are the vacancies, i.e., the vacuum [33] been, these solutions in equilibrium, in the case when one of the solutions is submerged in an external field.

Where the main supposition, made within the frame of this adsorption theory is that the role of the adsorption field could be simulated by; *Π.;* the so called "osmotic pressure" that is, the role of the energy of adsorption existing within zeolite or related materials channels and cavities, can be replicated by the pressure variation among the adsorbed and gas phases [32]; consequently, considering that the adsorption space is an inert volume; hence, the adsorption effect is produced by a virtual pressure compressing the adsorbed phase in this volume; in this case, we will have only a volume, i.e., a void adsorption space, in which an external pressure, *Π took* the role of the adsorption field [33]. Therefore, applying the hypothesizes of the osmotic theory of adsorption is feasible to affirm that the volume occupied by the adsorbate, V_a, and the vacancies, V_x, or free volume is [29] $V_a + V_x = V$; now in view of the fact that the volume occupied by an adsorbed molecule, *b*, and a vacancy is the same, therefore:

$$\frac{V_a}{b} + \frac{V_x}{b} = n_a + N^x = \frac{V}{b} = N_a$$

Consequently, if we multiply the previous equation by $\frac{1}{N_a}$, we obtain:

$$\frac{V_a}{N_a b} + \frac{V_x}{N_a b} = \frac{n_a}{N_a} + \frac{N^x}{N_a} = \frac{V}{N_a b} = X_a + X^x = 1$$

where X_a and. X^x correspondingly are the molar fractions of adsorbed molecules and vacancies. Then considering that the adsorption process in a micropore system can be described as an osmotic process; in which vacuum, that is, the vacancies is the solvent, whereas the adsorbed molecules is the solute [3]; hence, using the Osmosis Thermodynamics methodology applied to the above described model, it is possible to obtain the following adsorption isotherm equation [29, 32, 33]:

$$n_a = \frac{N_u K_0 P^B}{1 + K_0 P^B} \tag{13}$$

It is termed the osmotic isotherm of adsorption, or the Sips, or Bradleys isotherm equation, where this isotherm equation, fairly well, describes the experimental data of adsorption in zeolites, and other micoroporous materials; the linear form of the osmotic equation being expressed as follows:

$$y = P^B = N_a \left(\frac{P^B}{n_a}\right) + \frac{1}{K} = mx + b \tag{14}$$

where $y = P^B$, $x = \frac{P^B}{n_a}$, $m = N_a$, is the slope and $b = \frac{1}{K}$ is the intercept. Now it is necessary to state that for B = 1 the Osmotic isotherm reduces to a Langmuir-type adsorption isotherm for volume filling:

$$n_a = \frac{N_a K_{0,L} P}{1 + K_{0,L}, P}$$

In **Figure 6**, the linear plot of Eq. (14) is shown, using B = 0.5, fitting NH_3 at 300 K adsorption results in Mg-CMT, i.e., homoionic magnesium natural zeolite,

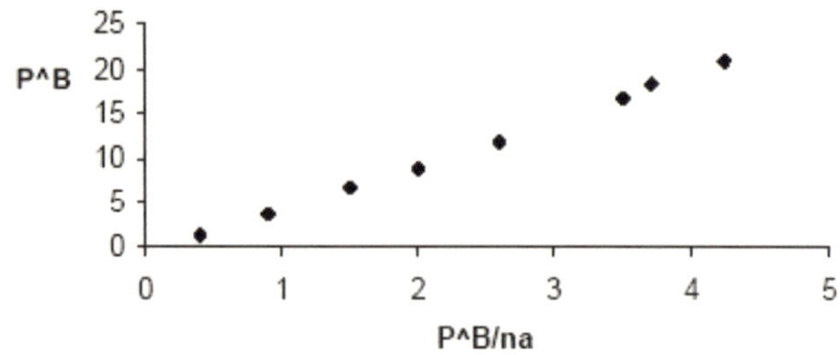

Figure 6.
Linear osmotic plot, with B = 0.5 of the adsorption data of NH_3 at 300 K in magnesium homoionic CMT zeolite [32].

concretely a blend of mordenite (39 wt. %), clinoptilolite (42 wt. %) along with additional phases (15 wt. %), where these supplementary phases are: montmorillonite (2–10 wt. %), calcite (1–6 wt. %), feldspars (0–1 wt. %), volcanic glass together with quartz (1–5 wt. %). These results were measured in a Pyrex glass volumetric adsorption vacuum system, consisting of sample holder, dead volume, dose volume, U-tube manometer, and thermostat [32]; this plot allowed us to calculate the maximum adsorption capacity of this zeolite, which is $m = N_a = 5.07$ mmol/g, and $b = \frac{1}{K} = -0.92$, $[(\text{Torr})^{0.5}]$; as a conclusion, it is possible to affirm that the experimental data are correctly fitted by Eq. (14).

The fitting process of the osmotic isotherm equation could be also carried out with the help of a non-linear regression method, where the fitting process allows us to calculate the best fitting parameters of the Eq. (13), i.e., N_a, K_O, and B, if this parameter is not taken as a constant, for example, $B = 1$; besides the program calculates the regression coefficient and the standard errors.

Further adsorption isotherms are reported below [35] (**Figures** 7 and **8**):

Additionally, the adsorption of oxygen (O_2) and nitrogen (N_2) in modified natural mordenite from the Palmarito, Santiago de Cuba, Cuba, deposit, composed of mordenite (80 wt.%), clinoptilolite (5 wt.%), and other phases (15 wt.%), where the other phases included montmorillonite (2–10 wt.%), quartz (1–5 wt.%), calcite (1–6 wt.%), feldspars (0–1 wt.%), and volcanic glass [2]. Labeled MP are reported in **Table 1**, where H means acid; NH, ammonia; Li, Lithium; Na, Sodium; K, Potassium; Mg, Magnesium; Ca, Calcium; Sr., Strontium; Ba, Barium zeolite; where, the oxygen nitrogen selectivity is given by [36].

$$\alpha = Y_a X_a / Y_g X_g$$

where X_a *and* Ya are the molar fractions of the adsorbates in the adsorbed phase, while X_g and Y_g are the molar fractions of both adsorbates in the gas phase.

It is necessary now to state that the framework of mordenite shows an orthorhombic unit cell displaying the space group Cmcm or Cmc21; in this sese, the unit cell of the Na cationic form has dimensions a = 18.13 Å, b = 20.49 Å, and c = 7.52 Å,

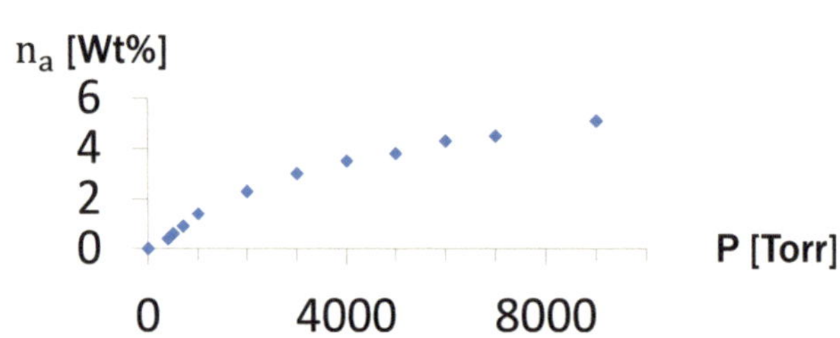

Figure 7.
Nitrogen adsorption on zeolite 5 a (Ca, Na-LTA) at 291 K.

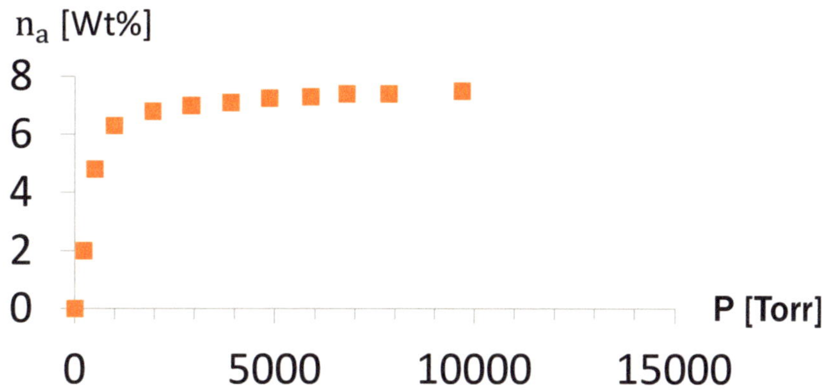

Figure 8.
Ethane adsorption on zeolite 13 X (Na-FAU) at 291 K.

ZEOLITES	$N_{MAX.OXYGEN}$ (mmol/g)	$N_{MAX.NITROGEN}$ (mmol/g)	K_{OXYGEN} $(Torr)^{-1} \times 10^{-3}$	$K_{NITROGEN}$ $(Torr)^{-11} \times 10^{-3}$	$\alpha_{NITROG-\ OXIG}$
H-MP	0.24	0.5	1.3	2.3	3.7
NH-MP	0.10	0.25	0.12	2.0	42
Li-MP	0.21	0.86	1.8	3.0	6.8
Na-MP	0.29	0.63	9.3	2.0	4.7
K-MP	0.07	0.26	3.1	3.7	4.4
Mg-MP	0.13	0.56	0.21	3.9	84
Ca-MP	0.27	0.56	1.4	4.8	7.1
Sr-MP	0.32	0.59	1.2	4.6	7.1
Ba-MP	0.23	0.75	1.5	1.9	4.1

Table 1.
Maximum adsorption oxygen ($N_{MAX.OXIGENO}$) and nitrogen ($N_{MAX.NITROGENO}$), Langmuir constants (K_{OXYGEN} and $K_{NITROGEN}$) and oxygen- nitrogen selectivity ($\alpha_{NITROG-\ OXIG}$) on modified natural mordenite MP [36].

along with the following composition: $Na_8Al_8Si_{40}\ O_{96} \cdot 24H_2O$; composed fundamentally of a channel system composed of a 5-membered ring system parallel to [001], having a free diameter of 6.6 Å, interconnected by smaller channels, parallel to [010], of 2.8 Å free diameter. However, the existence of stacking faults in the framework reduces the effective diameter of the channels to about 4 Å [37]. Moreover, the volume of the unit cell is $2794°A3$, while the kinetic diameters of nitrogen and oxygen are $d_N = 3.6$ Å and $d_O = 3.5$ Å. On the other hand, the quadrupole interaction (Q) of nitrogen is larger than that of oxygen, i.e., $Q_N = 0.31$ Debye > $Q_O = 0.10$ Debye; meanwhile, the polarizability (P) of nitrogen is also bigger than that of oxygen, that is, $P_N = 4.31°A3 > P_O = 3.96°A3$, facts that explain the higher values measured for the Langmuir constants in the case of nitrogen in comparison with those measured for oxygen adsorption [36].

The application of the grand canonical ensemble (GCE) allows us to handle the adsorption process in microporous materials such as zeolites and related materials. In this case, the whole zeolite is considered a GCE, i.e., the zeolite cavities or channels are considered in the frame of this model as independent open systems constituting the ensemble, additionally the adsorption field within the cavities is energetically homogeneous, i.e., the adsorption field is the same at any site within

the adsorption space; besides, each cage can accommodate$m = w/b$, molecules, where w and b are the volumes of the cavity and the adsorbed molecule, respectively. Hence if the ensemble is constituted by M independent open cavities, i.e., systems, the grand canonical partition function of the zeolite is given by [32]:
where

$$\Theta = \left[1 + \lambda Z(1) + \lambda^2 Z(2) + \ldots\ldots + \lambda^n Z(n)\right]^M = \overline{Z}^M \tag{15}$$

$$\overline{Z} = \sum_{N=0}^{m} \lambda^N Z(N) \tag{16}$$

Is the channel or cavity grand canonical partition function; been, *Z(N)* the Canonical Partition Function for *N* molecules in the channel or cavity ($0 < N < m$), while the absolute activity is given by $\lambda = \exp(\mu/RT)$ while, μ *is* the Chemical Potential. Thereafter, as:

$$\overline{N} = \frac{\partial \ln \Theta}{\partial \lambda} = RT\left(\frac{\partial \ln \Theta}{\partial \mu}\right) \tag{17}$$

$$\theta = \frac{\overline{N}}{mM} = \frac{M\overline{N}\,\overline{N}}{Mm\,m}$$

Now it is necessary to state that within the frame of the adsorption process in zeolites and related materials two cases are possible; that is, inmobile (I) or mobile (M) adsorption, been in the first and second cases, the canonical partition functions for the inmobile (Z_i) and mobile (Z_M) cases, in a homogeneous field without lateral interactions between the adsorbed molecules for $N < m$ given by:

$$Z_I(N) = \frac{m!}{N!(m-N)!}\left(Z_a^i\right)^N \exp\left(-\frac{N\left(E_0^a + \eta E_i\right)}{RT}\right) \tag{18}$$

$$Z_M(N) = \frac{w^N}{N!}(\Lambda)^N\left(Z_a^i\right)^N \exp\left(-N\left[\frac{\left(E_0 + \alpha(N/w)_i\right]}{RT}\right)$$

Now since:

$$\overline{N} = \frac{\partial \ln \overline{Z}}{\partial \lambda} = \lambda\left(\frac{\partial \ln \overline{Z}}{\partial \lambda} = \frac{A}{B}\right)$$

Consequently:

$$\theta = \frac{\overline{N}}{m} = \frac{K_I P}{1 + K_I P} \tag{19}$$

$$\theta = \frac{\overline{N}}{m} = \frac{K_M P}{1 + K_{MI} P}$$

With:

$$K_I = \left[\frac{Z^i}{Z_g^i}\right]\left(\frac{1}{RT\Lambda}\right)\exp\left\{\frac{\left[E_0^g - E_0^a\right] + \Omega\theta}{RT}\right\} \tag{20}$$

$$K_M = \left[\frac{Z^i}{Z_g^i}\right]\left(\frac{b}{RT}\right)\exp\left\{\frac{\left[E_0^g - E_0^a\right] + \Phi\theta}{RT}\right\}$$

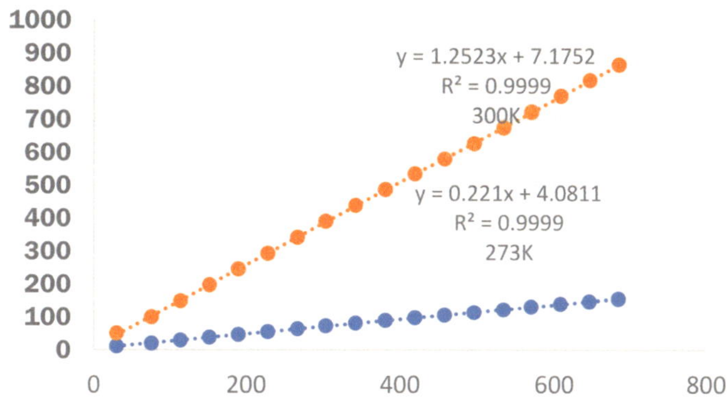

Figure 9.
Carbon dioxide adsorption on Ni-NP at 273 and 300 K [34].

Equation (19) is of the Fowler-Guggenheim type (FGT), but describing volume filling rather than surface covering, where both equations reduce to Langmuir-type (LT) isotherm equations, as well describing volume filling. Moreover, the osmotic isotherm equation and the FGT types have the same mathematical form in the case where B = 1 in the osmotic equation, as well as are equivalent to the Langmuir T type equations, when $\Omega = 0; \Phi = 0$.

To test these isotherm types, the FGT and LT type equations can be written as follows [2]:

$$\ln\left(\frac{\theta}{1-\theta}\right) = LnK + \frac{k\theta}{RT} \tag{21}$$

While the linear form of the LT type isotherms is $P = N_a(P/n_a) + 1/K = ax + b$. In **Figure 9**, the plot of carbon dioxide adsorption on Ni-NP at 273 and 300 K is reported [34].

8. t-plot method

The t-plot method suppose that the adsorbed phase as a liquid adhered film over the solid surface, the model was proposed by Halsey, De Boer, and coworkers following ideas previously proposed by Frenkel-Halsey-&-Hill which stated that it is possible to calculate, t, the width in (Angstrom) of the adsorbed layer, or multilayer thickness [32] and plot it as a function of: $x = P/P_o$. The methodology is effective for a multilayer adsorption; in which, the surface liquid film is supposed to show a unvarying width, t, density equal to the bulk liquid adsorbate, ρ_L, and be in contact with a uniform surface that produces an attraction adsorption field over the solid surface; hence, based on the aforementioned arguments, the adsorbed amount will be [38]:

$$n_a = \rho_L t$$

Attraction by the adsorption field is given by $V(z) = \frac{A}{z^9} - \frac{B}{z^m}$. Thereafter, following a model similar to that proposed by Polanyi, where it is considered that the entropy contribution to the free energy is small in comparison with the change of enthalpy, we have $\mu - \mu_L = RT\ln\left(\frac{P}{P_0}\right)$. Hence, supposing now that [38]:

$$\mu - \mu_L = V(z) \tag{22}$$

Subsequently, since the adsorption process considered in the present model is a multilayer one, then $V(z) \approx -\frac{B}{z^m}$ and consequently $RT \ln\left(\frac{P}{P_0}\right) = -\frac{B}{z^m} = -\frac{C}{t^m}$; hence, the thickness could be evaluated after normalizing an adsorption isotherm for an adsorbent that does not possess micropores, or mesopores. Subsequently, the multiplayer thickness, t, can be calculated by the following relation:

$$t = \frac{n_a}{N_m} d_0$$

where d_O is the thickness of a monolayer.

Thereafter, supposing again, that the surface liquid film is assumed of uniform width, t, along with having a density equal to the bulk liquid adsorbate, ρ_L, hence, we have [32]:

$$d_0 = \frac{M}{\sigma N_A \rho_L}$$

where N_A is the number of Avogado and σ is the cross-sectional area, that is, the normal area that each molecule occupies in a completed monolayer. For instance, if $\sigma(N_2)$ = 0.162 nm^2 for N_2 at 77 K, $M(N_2)$ = 28.1 g/mol, and $\rho_L(N_2)$ = 0.809 g/cm^3 then, d_O = 0.354 nm.

The following relation *t versus, (1/x), is used between others* to carry out the t-plot [32, 38]:

$$t = 3.54\left(\frac{5}{2.303 \log\left(\frac{P_0}{P}\right)}\right)^{1/3} \tag{23}$$

which is the Halsey equation, valid for N_2 at 77 K or the equation used by De Boer. Moreover, the following expression:

$$t = \left(\frac{13.9}{0.034 + \log\left(\frac{P_0}{P}\right)}\right)^{1/2}$$

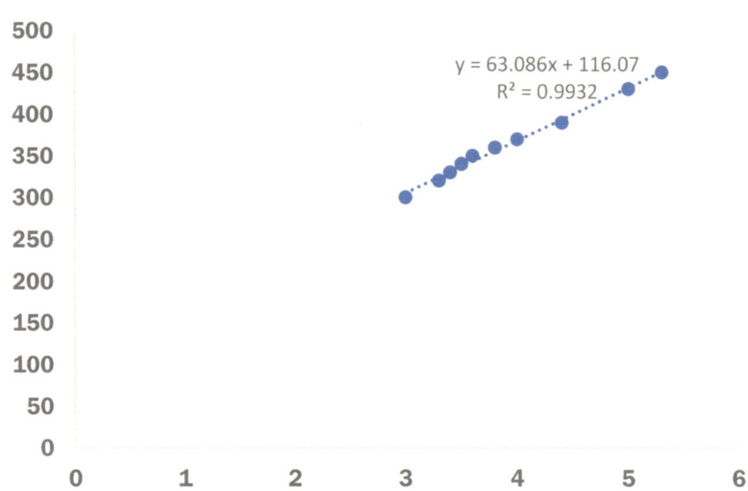

Figure 10.
t-plot for the adsorption of N_2 at 77 K in a high specific surface silica labeled 70bs-2-25C [39].

is as well used for the application of the t-plot methodology. Now it is necessary to state that the molar volume of liquid nitrogen at 77 K is $V(77K)_M^{N_2} = 32.565$ (cm^3/mol) [37].

The method to calculate, W, the microporous volume [in (cm^3/g)] and S [in (m^2/g)] the outer surface applying the t–plot method is as follows: after the elimination of the points that do not fit a linear plot (**Figure 10**):

$$y = n_a = Rt + N_a = mx + b$$

The linear regression is made, then the intercept, $b = N_a$, and the slope, $m = R$, are calculated, given that the intercept is related to the micropore volume, W^{MP}; thereafter, using the Gurvich rule $W^{MP} = N_a V_L$, where V_L, ($V(77K)_M^{N_2}$=32.565 $\frac{cm^3}{mol}$) is the molar volume of the adsorptive at the adsorption temperature, T, [37] is then calculated micropore volume; $W^{MP} = 116.1x32.565 = 5.17 mmol/g_L$; together with, the outer surface, $S = RV_L = 63.08x32.565 = 2,054 m^2/g$ [39].

9. Thermodynamics of adsorption

Adsorption is a general tendency of matter, and during its occurrence, a decrease in the surface tension is experienced by the solid. For this reason, adsorption is a spontaneous process, where the Gibbs free energy decreases, i.e., $\Delta G < 0$. Besides, in the course of physical adsorption, molecules from a chaotic bulk phase are transferred to a relatively ordered adsorbed state, since in the adsorbed state molecules can only move within the surface or a pore; consequently, in the course of adsorption by the entire system, a reduction of entropy takes place, i.e., $\Delta S < 0$; subsequently [2, 32]:

$$\begin{aligned} &\Delta G = \Delta H - T\Delta S \\ &or \\ &\Delta G = \Delta H - T\Delta S < 0 \end{aligned} \tag{24}$$

Consequently, the adsorption process releases heat, i.e., it is an exothermic process. Consequently, it is stimulated by a reduction of the adsorption experiment temperature.

As a matter of fact, the main thermodynamic relation for a bulk mixture system is given by:

$$dU = TdS - PdV + \sum_i \mu_i dn_i \tag{25}$$

where S is their entropy, U is the internal energy, V is the volume, T is the temperature, μ_i is the chemical potential, and n_i is the constituent number of moles included in the system [18]. Now it is supposed that the adsorbent together with the adsorbed gas is a solid solution, labeled, system aA; thereafter, using the proposed scheme, it is possible to deduce the following equation, which thermodynamically describes the aA system [2, 32]:

$$dU_{aA} = TdS_{aA} - PdV_{aA} + \mu_A dn_A + \mu_a dn_a$$

where U_{aA}, S_{aA}, and V_{aA} are the internal energy, entropy, and volume of the system aA, respectively, and μ_a and μ_A are the chemical potentials of the adsorbate, a, and the adsorbent, A, while n_a and n_A are the number of moles of the adsorbate and the adsorbent in the system aA, respectively. Now if we define $\Gamma = n_a/n_A$; then:

$$\mu_a = \mu_a(T, P, \Gamma) \text{ and } \mu_A = \mu_A(T, P) \tag{26}$$

As a result:

$$\left[\frac{d \ln P}{dT}\right]_\Gamma = \frac{\overline{H}_g - \overline{H}_a}{RT^2} = \frac{q_{iso}}{RT^2} \tag{27}$$

where H_g and H_a are the partial molar enthalpies of the adsorbate in the gas phase and in the aA system, respectively; now, applying Eq. (16), it is possible to define the isosteric enthalpy of adsorption [31]

$$q_{iso} = \overline{H}_g - \overline{H}_a \tag{28}$$

where q_{iso} is the enthalpy of desorption, or the isosteric heat of adsorption, which is calculated with the help of adsorption isotherms. An additional significant adsorption heat is the differential heat of adsorption, since when an adsorbate contacts an adsorbent, heat is released. In our case, the thermal effect produced was measured with the help of a thermocouple placed inside the adsorbent and referred at room temperature (**Figure 11**), a variety of the Tian–Calvet heat-flow calorimeter [32]. This calorimetric methodology is characterized by the fact that the difference of temperature among the tested adsorbent and a thermostat is determined; therefore, in this heat-flow calorimeter, the produced thermal energy in the adsorption cell is permitted to flow with no limitations to the thermostat. In the calorimeter constructed by us, heat flows throughout a thermocouple; thereafter, the voltage produced by the thermocouple, which is proportional to the thermal power, is amplified and recorded in an x-y plotter (see **Figure 11**); in which, the actual thermal effect generated is the integral heat of adsorption, measured using the Eq. (2)

$$\Delta Q = k \int_0^{t_{max}} \Delta T dt \tag{29}$$

where ΔQ is the integral heat of adsorption released during the finite increment, κ is a calibration constant, ΔT is the difference between thermostat temperature and the sample temperature during adsorption, and t is time, the differential heat of adsorption being calculated as follows [32, 33]:

$$q_{diff} = \frac{\Delta Q}{\Delta n_a} = \frac{k \int_0^{t_{max}} \Delta T dt}{\Delta n_a}$$

The heat-flow calorimeter used consisted of the high vacuum line for adsorption measurements applying the volumetric method; as reported in **Figure 11**, the equipment comprises the following: a Pyrex glass, vacuum system including a sample holder, a dead volume, a dose volume, a U-tube manometer, and a thermostat, including now a thermocouple immersed in the adsorbent bed, which was coupled to an x-y plotter.

In **Figure 12**, the measurement of the differential heat of adsorption for the adsorption of carbon dioxide in the natural mordenite labeled MP is reported, which is a mordenite from the Palmarito, Santiago de Cuba, Cuba, deposit (mordenite (80 wt.%), clinoptilolite (5 wt.%), together with montmorillonite (2–10 wt.%), quartz (1–5 wt.%), calcite (1–6 wt.%), feldspars (0–1 wt.%), and volcanic glass) [2]. In **Figure 12**, the obtained data are shown indicating that CO_2 adsorption process in this zeolite is energetically heterogeneous; i.e., the heat of adsorption is a diminishing function of the zeolite micropore volume recovery, i.e., $\theta = n_a/n_{max}$. That is, the plot of q_{diff} versus θ shows the following: two steps, one at 90 kJ/mol and the other at 70 kJ/mol, where was released, moderately high values of the adsorption heats, indicating that CO_2 molecules powerfully interact through their quadrupole moments with the mordenite framework; after that, a reduction of

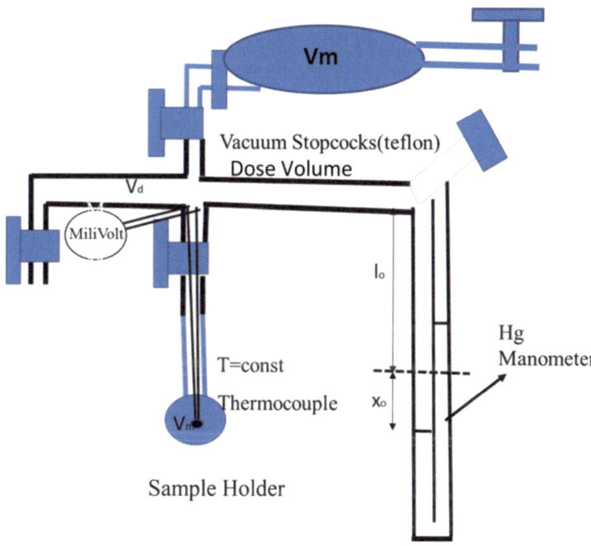

Figure 11.
Adsorption volumetric equipment.

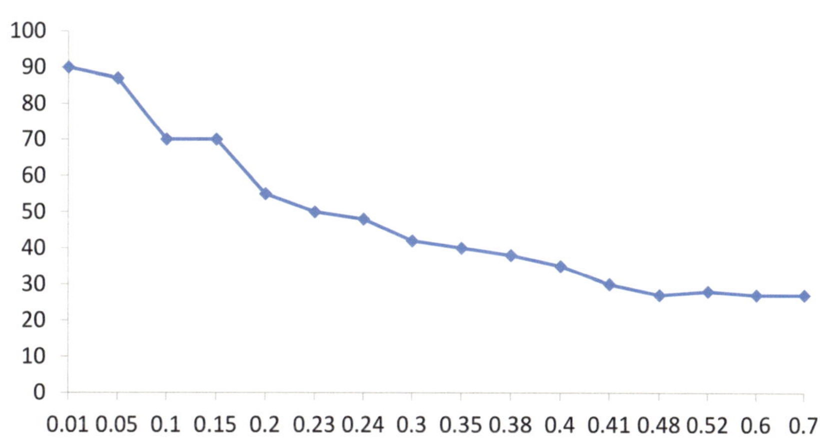

Figure 12.
Plot of q_{diff} vs. θ for the adsorption of CO_2 at 300 K in the MP zeolite [2].

the adsorption heat is found, up to a value corresponding to the bulk heat of condensation of the CO_2 molecules [26–28].

10. Adsorption interaction fields

If a molecule contacts a solid adsorbent, it is exposed to various interaction fields, such as dispersion attraction, φ_D; repulsion, φ_R; induced polarization, φ_P; permanent dipole, $\varphi_{E\mu}$; quadrupole, φ_{EQ}; and sorbate-sorbate, φ_{AA}.

Specific interactions as the acid-base interaction with the active site, φ_{AB}, if the surface contains hydroxyl bridge groups.

The dispersion or London forces among adsorbed nonpolar molecules and all types of adsorbents takes place when the transient dipoles turn out to be correlated; then, the dipoles of the nonpolar adsorbed species prompt a dipole in the atoms of the adsorbed species that act together to reduce the energy of the adsorption system. Hence, due to the correlation, the prompted dipoles created in the entire arrangement do not disappear, generating a dipole–dipole collaboration, dispersion interaction being intensity dependent on the polarizability of the adsorbate molecule and the adsorbent surface atom [40].

Additionally, the electrostatic contribution to the potential includes the dipole induced, dipole permanent and quadrupole terms, where, the induced polarization term occurs when nonpolar molecules within an electric field are polarized, and then an induced dipole moment is produced, the permanent dipole and quadrupole terms being caused by molecules whose structures produce permanent dipoles and quadrupoles; for example, H_2O, H_2S, SO_2, and NH_3 are molecules with a high dipole moment, whereas CO_2 is a molecule with a high quadrupole moment [1].

The electrostatic attractive interactions are stronger than the dispersion interactions; however, dispersion is the fundamental attractive force present during adsorption in all adsorbate-adsorbent systems, for example, in the case of molecules like H_2, Ar, CH_4, N_2, and O_2. Given that the dipole moments of these molecules are zero, the quadrupole moment is very low or absent, and the polarization effect will only be noticeable in the case of adsorbents with high electric fields; the dispersion and repulsion interactions are responsible for the adsorption effect present in all adsorption gas-solid systems; therefore, they are nonspecific interactions [2].

Author details

Rolando Roque-Malherbe[1] and Carlos de las Pozas del Rio[2]*

1 BioTECNOS Limited, Talca, Chile-San Sebastian University, Santiago, Chile

2 University San Sebastian, Santiago, Chile

*Address all correspondence to: carlos.delaspozas@uss.cl

References

[1] Breck DW. Zeolite Molecular Sieves. New York: John Wiley & Sons; 1974. pp. 635-700

[2] Roque-Malherbe R. Physical Chemistry of Materials. Energy and Environmental Applications. Boca Raton Florida: CRC Press; 2009. pp. 275-338

[3] Roque-Malherbe R. Adsorption and Diffusion in Nanoporous Materials. Boca Raton, FL, USA: CRC Press; 2018. pp. 153-287

[4] Benitez F, Travieso L, Sanchez E, Roque-Malherbe R. Waste water treatment technology. In: International Conference on Advanced Waste Water Treatment and Reclamation; September 1989; Cracow, Poland

[5] Misaelides P. Application of natural zeolites in environmental remediation. Microporous and Mesoporous Materials. 2011;**144**:15-18

[6] Wang S, Peng Y. Chemical Engineering Journal. 2010;**156**:11-24

[7] McKeen JC. Proton and ion conduction in microporous materials [Ph. D. Thesis]. Pasadena, CA, USA: California Institute of Technology; 2009

[8] Zagorodni AA. Ion Exchange Materials: Properties and Applications. Amsterdam, the Netherlands: Elsevier Science & Technology Books; 2006

[9] Lu C. Nanoporous materials chemistry. From gas separation to high pressure chemistry [Ph. D. Thesis]. Bethlehem, PA, USA: Lehigh University; 2017

[10] Caro J, Noack M. Zeolite membranes–recent developments and progress. Microporus and Mesoporous Materials. 2008;**115**:215-233

[11] Thomas JM, Thomas WJ. Principle and Practice of Heterogeneous Catalysis. New York: VCH Publishers; 1997

[12] Schlögl R. In: Ertl G, Knozinger H, Weitkamp J, editors. Preparation of Solid Catalysts. Weinheim, Germany: WileyVCH; 1999. pp. 11-28

[13] Thomas TL. Process for cation separation using zeolite materials. U. S. Patent 3,033,641; 1962

[14] McNaught AD, Wilkinson A. IUPAC Compendium of Chemical Technology. Oxford: Blackwell Scientific Publications; 1997

[15] Do JL, Friscic T. ACS Central Science. 2017;**3**:13-19

[16] Sanchez DA, Ortega N, Kumar A, Roque-Malherbe R, Polanco R, Katiyar RS, et al. Symmetries and multiferroic properties of novel room-temperature magnetoelectrics: Lead iron tantalate–Lead zirconate titanate (PFT/PZT). AIP Advances 2011;**1**:042169-1-042169-14

[17] Brinker C, Scherrer G. Introduction. In: Brinker C, Scherrer G, editors. Sol-Gel Science. 1st ed. The Physics and Chemistry of Sol-Gel Processing. New York: Academic Press; 2013. pp. 1-18

[18] Robson H, editor. Verified Synthesis of Zeolitic Materials. Amsterdam, the Netherlands: Elsevier; 2001

[19] de las Pozas C, Diaz Quintanilla D, Perez Pariente J, Roque Malherbe R, Magi M. Hydrothermal transformation of natural clinoptilolite to zeolites Y and P1: Influence of the Na, K content. Zeolites Zeolites. 1989;**9**:33-39

[20] Martens JA, Jacobs PA. Synthesis of High Silica Aluminosilicate Zeolites. Vol. 33. Amsterdam: Elsevier; 1987

[21] Gonzales E, Lariot C, Villegas R, Morales F, Cañizares H, Herrero E, et al. Influence of temperatura and syntheisis time in the acidity and morphology of ZSM-5 zeolites. Revista CENIC. 2001;**32**: 43-50

[22] Roque-Malherbe R, Lopez-Cordero R, Gonzales-Morales JA, Oñate-Martinez J, Carreras-Gracial M. A comparative study of MeAPO molecular sieves with AFI structure type. Zeolites. 1993;**13**:481-484

[23] Marsh H, Rodriguez Reynoso F. Active Carbon. Amsterdam: Elsevier; 2006

[24] Bansal RC, Gayal M. Activated Carbon Adsorption. Boca Raton, FL, USA: CRC Press-Taylor and Francis; 2005

[25] Roque-Malherbe R. Physical Chemistry of Materials. Energy and Environmental Applications. Boca Raton Florida: CRC Press; 2009. pp. 103-135

[26] Roque-Malherbe R, Lugo F, Polanco R. Applied Surface Science. 2016;**385**: 360-367

[27] Roque-Malherbe R, Carballo E, Polanco R, Lugo F, Lozano C. Structure and adsorption properties of a porous cooper hexacyanoferrate polymorph. Journal of Physics and Chemistry of Solids. 2015;**86**:65-73

[28] Roque-Malherbe R, Uwakweh ONC, Lozano C, Polanco R, Hernandez-Maldonado A, Fierro P, et al. Structural effects and interactions of carbon dioxide molecules adsorbed on Ni, Zn, and Cd nitroprussides. Journal of Physical Chemistry C. 2011;**115**: 15555-15569

[29] Roque-Malherbe R. Physical Chemistry of Materials. Energy and Environmental Applications. Boca Raton Florida: CRC Press; 2009. pp. 137-218

[30] Reguera E, delas Pozas C, Diaz Aguila C, Roque Malherbe R. Mössbauer study of hydrothermal transformation of natural clinoptilolite into Y and P1 zeolites. Journal of Solid State Chemistry. 1991;**94**:215-219

[31] Everett DH, Ottewill RH. Surface Area Determination, International Union of Pure and Applied Chemistry. London: Butterworth; 2103

[32] Roque-Malherbe R. Adsorption and Diffusion in Nanoporous Materials. Boca Raton, FL, USA: CRC Press; 2018. pp. 35-71

[33] Bering BP, Dubinin MM, Serpinskii VV. On thermodynamics of adsorption in micropores. Journal of Colloid and Interface Science. 1972;**38**:185-194

[34] Roque-Malherbe R, Lozano C, Polanco R, Marquez F, Lugo F, Hernandez-Maldonado A, et al. Study of carbon dioxide adsorption on a Cu-nitroprusside polymorph. Journal of Solid State Chemistry. 2011;**184**(5): 1236-1244

[35] Venero A, C and L Science and Technology, Hiahleah, FL, USA, Personal Communications

[36] Autie M, de las Pozas C, Balmaceda J. Revista CENIC. Ciencias Químicas. 2003;**34**:41-44

[37] Lide RD, editor. CRC Handbook of Chemistry and Physics. 88th ed. Boca Raton, Florida: Taylor & Francis Group; 2008

[38] Voogd J, Scholten JJF, Van Bekkum H. Nitrogen adsorption isotherms at 77 K have been measured of H-ZSM-5 samples. Colloids and Surfaces. 1991;**55**: 163-171

[39] Marquez-Linares F, Roque-Malherbe R. Journal of Nanoscience and Nanotechnology. 2006;**6**:1-5

[40] Roque-Malherbe R. Adsorption and Diffusion in Nanoporous Materials. Boca Raton, FL, USA: CRC Press; 2018. pp. 35-48

Chapter 5

Composition Electrolytic Coatings with Given Functional Properties

Gulmira Yar-Mukhamedova, Maryna Ved',
Nikolay Sakhnenko and Tetiana Nenastina

Abstract

Ternary alloys of cobalt with molybdenum and tungsten deposited from biligand citrate-pyrophosphate electrolyte by pulsed mode exhibit different compositions and surface morphologies depending on current density and on/off time. The structure of binary and ternary alloys was found to be amorphous crystalline, and intermetallic phases Co_7W_6 and Co_7Mo_3 were identified in deposits. The coherent-scattering region size of the amorphous part was detected of 2–8 nm. The amorphous structure of ternary alloys and significant content of alloying elements (Mo and W) predetermine improved high corrosion resistance. Corrosion resistance of binary and ternary deposits increases with total content of refractory metals, which associated with molybdenum and tungsten, enhancing corrosion resistance to pitting as well as decreasing in roughness and smoothing out the relief of ternary coatings. Ternary galvanic alloys of cobalt with molybdenum and zirconium with micro-globular morphology with low level of stress and cracks are formed at a current density of 4–6 A dm^{-2} and polarization on/off time 2/10 ms. High corrosion resistance of ternary coatings based on cobalt is caused by the increased tendency to passivity and high resistance to pitting corrosion in the presence of molybdenum and zirconium, as well as the acid nature of their oxides.

Keywords: coating, electrodeposition, composition, physicomechanical properties

1. Introduction

The range of functional coatings with alloys, the creation of which to a large extent predetermined the progress of modern technologies, can be fully attributed to the advances of electrochemical science. Quite obvious consumer advantages of coatings with alloys, in comparison with monometallic analogues, led to prospects for their widespread use, despite a number of technical difficulties encountered. The twenty-first century brought new challenges—amorphous alloys, nanoscale, nanocrystalline, and nanolaminate structures, film materials with gigantic magnetic resistance or high-temperature superconductivity, multiferroics, etc., became the order of the times, the formation of which on the basis of trivial bimetallic compositions proved impossible. This led to the creation of a new paradigm in practical electroplating—the only alternative to the transition to multicomponent and synergistic alloys and composites. The implementation of it put a number of pressing problems in the design of such polymetallic systems.

IntechOpen

The essence of the problem lies in the absence of clear design algorithms for multicomponent alloys with a given level of properties and the presence of a significant number of empirical views, the use of which allowed electrochemical gurus to bypass the obstacles of Mother Nature in creating electroplated coatings with alloys. New tasks required new approaches, but before proceeding to their consideration, let us analyze the essence of the problem in the formulation "Galvanic alloys—the philosophy of synergism".

1.1 Alloys

The physical encyclopedia gives the classical definition of an alloy as "a metallic, macroscopically homogeneous system consisting of two or more metals (less commonly metals and nonmetals) with characteristic metallic properties." The etymology associates the material with the method of its preparation—the combination of individual components during melting followed by crystallization of the melt, although dozens of other methods are currently known. In addition, the definition of "macroscopically homogeneous" raises the question of whether heterogeneous structures (mechanical mixtures) belong to the alloy community. Apparently, it is more correct to speak of metal alloys as homogeneous or heterogeneous systems, as well as intermetallic compounds, although it is quite obvious that real alloys often contain all three types of these structures. Regarding the "metallic properties," we note that a wide class of compounds—"synthetic metals" [1, 2]—has all the above features (hardness, electrical and thermal conductivity, opacity, gloss, etc.), but does not contain metals at all, if you do not take attention cations.

1.2 Galvanic alloys

In this formulation, the nature of the material and the method of its production are inextricably linked, but with all the apparent uniqueness, and there is a number of significant differences not only in the structure and properties of metallurgical and galvanic alloys but also in the concentration ratios of the components in the material. During electrolytic deposition, alloys can be formed that differ essentially in their phase composition and properties from those obtained by thermal means. This expands the range of technical capabilities of electrolytic alloys and their field of application.

1.3 Synergism

The term "synergetic" means joint or corporate action, and the synergistic effect is an increase in the efficiency of activity as a result of integration, the merging of separate parts into a single system. The question is to what extent the terminology of synergetic, or rather, actually synergism, is applicable to such objects as electrolytic alloys? From the definition of the object of research of synergetic "processes in complex open non-equilibrium systems ..." [3], it follows that the combination of these system-forming features is fully applicable to process in electrochemical systems, which include alloy formation. Indeed, such an electrochemical system is complex, open, and nonequilibrium under the conditions of application external polarization.

For example, the advantage of coatings with alloys in comparison with individual metals, as well as the realization of synergism during the electrolytic alloy formation, manifests themselves in a change in the microhardness of materials depending on their composition and structure (**Table 1**). It is obvious that the formation of intermetallic compounds in the systems "metal of the iron family—molybdenum or tungsten," as well as the formation of amorphous structures, creates the prerequisites for a super-additive increase in the microhardness of the coatings.

Among the methods for solving complex technical problems, we will highlight both a deep understanding of the problem by creating a formalized description based on a systematic approach, followed by analysis and synthesis of complex systems, and a black box principle that does not involve a combination of the above actions, but is no less effective when using artificial intelligence tools, for example, artificial neural networks [9]. It is clear that the most significant results can be achieved with the integrated use of both approaches. In the first case, the electrochemical system (ECS), as a complex object in which the phenomena of substance transfer and electrical and thermal energy are realized, is characterized by a set of input and output variables, a set of parameters of electrode reactions and related physic-chemical transformations.

Since all components of the system are in constant development, they should be considered as processes, extending this procedure to subsystems. Therefore, to describe ECS in the field of external disturbances, it is necessary to identify the elements and links that make up the system and to establish the interaction between them, thus linking the functions and structures [10]. The set of processes in ECS can be represented as the implementation of transformations in the intersection of the subsets: ionic medium (electrolyte), electrode (electrode material), external disturbances (individual factors, as well as their joint or complex effect), as we have previously shown in the framework of corrosion protection analysis [11].

Among the most significant external influences, it is necessary to attribute, mainly, the different nature of the field (vector quantities), among which are:

- an electric field (polarization), including nonstationary modes of electrolysis with varying amplitude and temporal characteristics;

- a temperature field, which allows control not only the temperature of the reaction volume but also the aggregation state of individual elements (solutions-melts-gas phase), while the rate of change $\Delta T/\Delta t$ allows you to control the transformation routes (for example, supersaturation of solutions, the formation of 2- and 3-dimensional nuclei, the formation of amorphous structures, etc.);

- a magnetic field (factor in electrochemical processes is nonexhaustible, but the studied is completely insufficient),

- a pressure (creation of excessive pressure or vacuum);

Monometallic coatings	**H_V**	**Intermetallic compounds and alloys**	**H_V**
Cr (lustrous) [4]	7.5–11.0	CoW [6]	8.3
Cr (milk) [4]	4.5–6.0	Co_3W [6]	5.1
Cr (from tetrachromat bath) [4]	3.5–4.0	Co—W [5]	3–6.8
Co [5]	1.6	Fe—W [7]	13
Fe [5]	4.5–7	Ni—W [5]	2–14
W (from melt) [6]	3.7	Fe—W—P [8]	≥11

Table 1.
Microhardness (GPa) by Vickers galvanic coatings of metals and metal alloys.

- an impulse of movement, including the ultrasound field (mixing, transfer of reactants, or electrolysis products);

- a radiation (radiation field), which, depending on the nature, can lead, for example, to radiolysis, change of electrolysis conditions under the action of laser irradiation or structure and surface properties of the electrode material, as well as other changes in the state and properties of individual phases and even transformation routes;

- a gravitational field, which changes the conditions for the electrochemical processing (e.g., the stages of transportation), as well as other, less significant effects.

As an example, one can analyze the contribution of individual groups of process parameters (**Figure 1**) to the formation of synergistic alloys [12]. Thus, varying the composition of the electrolyte, the modes and parameters of the electrolysis, we electroplated iron alloys with refractory metals of different compositions and morphologies [13–15], which determine the level of functional properties (**Figure 2**).

Cobalt-based alloys are widely used as construction materials for different technic applications, for example, super alloys for aircraft turbine vanes and blades, alloys for powerful, high-coercive force magnets, hard metal alloys for cutting tool materials, and protective hard coatings. Cobalt is utilized as a matrix of special materials, including alloys for dental and surgical implants or bone fracture fixation, thermal resistant materials (Fe—Ni—Co, Co—Fe—Cr), magnetic recording thin films, catalysts [16], etc.

The addition of refractory metals (molybdenum, tungsten, zirconium, etc.) even in small amounts to cobalt significantly extends the functional properties of materials. In some cases, of particular interest is the implementing of these properties in thin surface layers. High adhesion and a wide range of coating composition are provided by electrodeposition from aqueous and nonaqueous solutions.

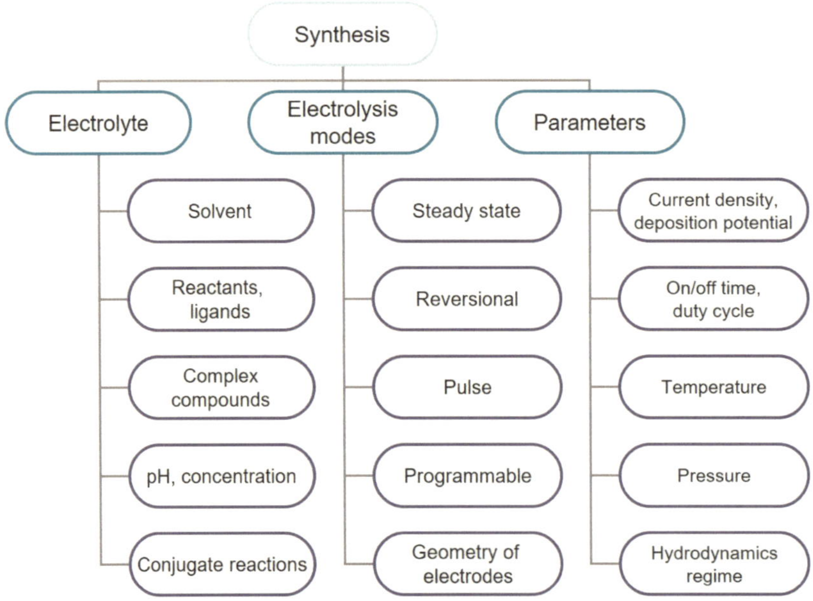

Figure 1.
The main factors of electrolytic alloying.

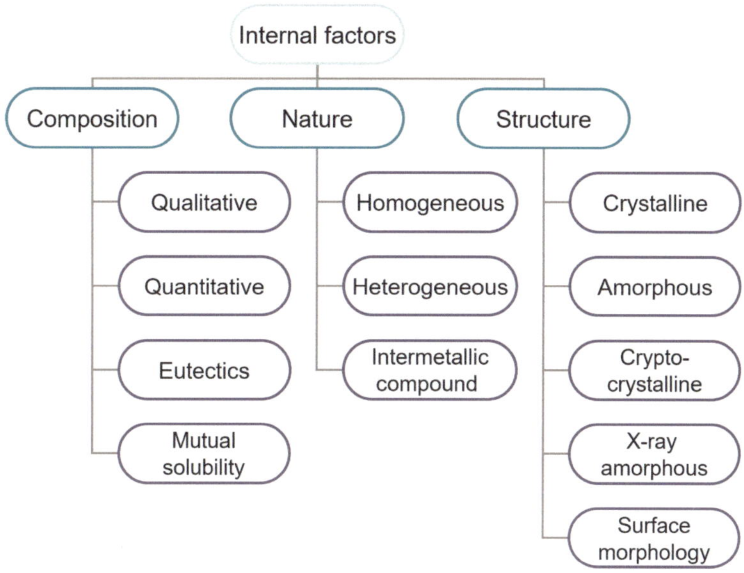

Figure 2.
Internal factors forming the functional properties of the alloys.

Electrodeposition peculiarities of binary and ternary cobalt alloys with molybdenum as well as their properties are reflected in promising publications. Cobalt-molybdenum (ω_W is 10%) films with low coercivity and high saturation magnetization were formed from a sulfate-citrate bath [17]. It was shown that the dependence of electrolytic alloy structure on the current density, namely a close-packed hexagonal structure, was formed at low cathodic polarization, and both crystalline and amorphous structures were appeared at higher polarization [18]. High adherent, compact, and uniform cobalt and cobalt-molybdenum coatings with 1–8 wt. % Mo were deposited onto copper substrates from ionic liquids based on choline chloride (ChCl) at a current density of 7–25 mA cm^{-2}, at 90–100°C [19]. Cobalt-molybdenum-boron amorphous electrolytic alloys of composition, wt. %: Co—51, Mo—47, and B—2, were deposited from citrate-phosphate-ammonia bath at a cathode current efficiency of 29–65%. Above materials exhibit high hardness, corrosion resistance, wear resistance, and also sufficient ductility [20]. Cobalt-molybdenum-phosphorus (Co—Mo—P) coatings containing 8% Mo, 20% P, and Co balance were deposited from citrate-phosphate electrolyte and were recommended as barrier layer to replace nickel [21]. The amorphous Co—Mo—C coatings were deposited with additional exposure of working electrode in the magnetic field. The content of Mo in the coatings deposited in a magnetic field increases up to 34.2 at. % as compared with traditionally deposited alloys [22]. A growth of overvoltage during hydrogen evolution at above materials was also observed.

Thus, in particular, the light lustrous uniform coatings made of iron binary and ternary alloys with refractory metals were deposited from complex electrolytes both at direct and pulse current. It was noted fairly high deposition rate up to 20 μm/h and the current efficiency of 60–85%, which is much higher as compared with the results of other scientists [23–25]. The morphology and topography of the coatings were shown to be depended on the nature of alloying refractory metals and the electrolysis modes on the course of the electrochemical alloy forming reactions. The reason was the change in the nature of the discharging particles and the limiting stage of the net electrode process.

The biligand citrate-pyrophosphate electrolyte [26, 27] was used for codeposition cobalt with molybdenum and tungsten to overcome significant potential difference of alloying metals which also are reduced multistage [12, 13]. The binary Co—Mo [26] and Co—W [5] deposits were obtained, and their composition was controlled by the variation of the pH and the concentration ratio of alloying metals' and ligands in a bath. The formation of hetero-nuclear complexes cobalt, molybdate, and tungstate with citrate and pyrophosphate, and complexes subsequently reducing into an alloy, may be competing with each other as it was shown for iron ternary [28, 29], Co—Mo—Zr [30] and Co—Mo—W deposits.

Ternary Co—Mo—Zr alloys were also deposited from citrate-pyrophosphate electrolyte with optimal concentration ratio of molybdate and tungstate for ligands [30, 31]. It was shown the advantage to use electrolyte with tungstate excess compared with molybdate. In addition, the use of pulse current was contributing to the deposition of ternary alloys with a high content of refractory metals.

2. Co—Mo—W electrolytic alloys

Figure 3 shows the current efficiency and composition of deposits Co—Mo—W plated from bi-ligand (citrate-pyrophosphate) bath in pulse regime at ratio t_{on}/t_{off} = 5/20 ms. The total deposition time is 30 min. As we can see from the plots, the total alloying metal content in the deposits Co—Mo—W rises from 19 at. % up to 30 at. % with increasing cathodic polarization from 4 to 10 A dm^{-2}. This phenomenon is quite natural and is due to the shift of the electrode potential in the site of oxometalates reducing with increasing current density. The decrease in the alloying metal content with an increase in the current density of more than 10 A dm^{-2} is associated with the intensification of the side reaction of hydrogen evolution. The current efficiency decreases with current density also due to the impact of the side reaction of hydrogen evolution. Indirectly, a decrease in the current efficiency with an increase in the refractory metal content may be due to the alloy catalytic activity in the electrolytic hydrogen evolution.

Both the morphology and composition of ternary alloys deposited in pulse mode at current density 8 A dm^{-2}, on/off time 5/20 ms depend on the electrolyte temperature (**Figure 4**). Since it was established earlier that lower current densities 5–7 A dm^{-2} favor higher current efficiency, this growth is due to the formation of

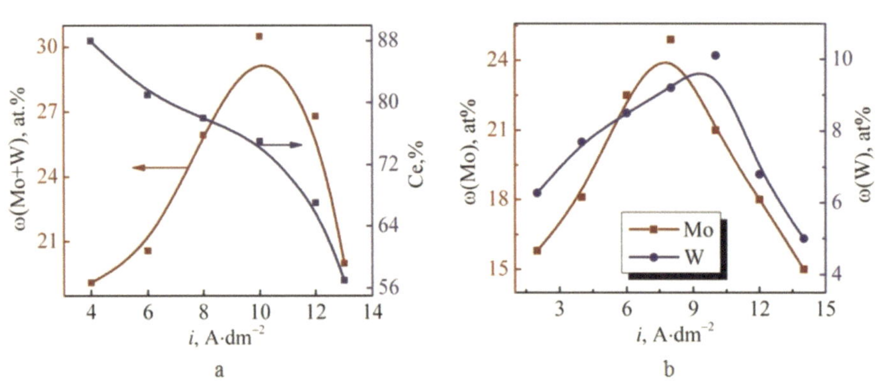

Figure 3.
Dependence of current efficiency Ce, total refractory metal content (a), and Mo or W content (b) in Co—Mo—W alloy on applied current density.

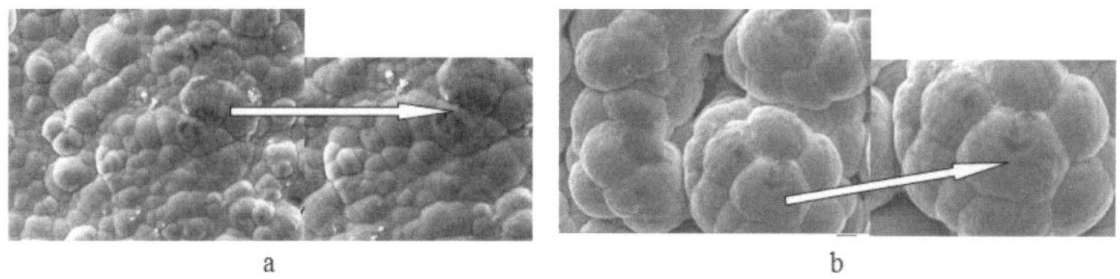

Figure 4.
Morphology and refractory metal content (at. % in terms of metal) in Co—Mo—W coatings, deposited at: (a) T = *30°C, Co—86.6, Mo—7.0, W—6.4; (b)* T = *50°C, Co—83.2, Mo—10.7, W—6.1. Magnification ×2000/5000.*

intermediate molybdenum and tungsten oxides instead of metals as suggested by the appearance of the deposited coatings. Those plated using current densities 8 A dm^{-2} are smooth, compact, and microglobular. Those obtained using higher temperature of 50°C (**Figure 4b**) have developed globular surface with larger sizes of agglomerates and microspheroids. With an increase in the temperature, the relative content of molybdenum also increases, and tungsten one changes slightly. Despite WO_4^{2-} : MoO_4^{2-} ratio in the electrolyte of 1:2, the coatings are enriched with molybdenum.

It is obviously molybdenum and tungsten compete with each other during deposition into an alloy, and therefore the atomic ratio of metals in the coating differs from the ratio of oxoanion concentrations in the electrolyte [30]. In turn, the atomic ratio of metals in the alloy determines the structure of the morphology of the surface layers. So, it is the increase in ω(W) that causes deep microcracks in the coating to a greater extent than the growth of the total refractory metal content [33].

Time parameters of pulsed electrolysis namely on/off time impact the current efficiency and composition of ternary alloys (**Figure 5**). Increasing the polarization time within the range of 2–5 ms at the current density of 10 A dm^{-2} and pause time 20 ms promotes content, both refractory components, in the coating (**Figure 5a**) but more significantly for tungsten. No any significant change in refractory metal content observed at longer polarization. As it follows from experiment (**Figure 5b**), the optimal off-time for alloying metal content is in the range of 15–20 ms when on-time is 5 ms. Thus, we can conclude that the pulse/pause ratio of 1/ (3–4) provides the maximum content of molybdenum and tungsten in the alloy. Considering the molybdenum competition with tungsten when forming hetero-nuclear complexes discharged at the cathode, the energy and time parameters of electrolysis despite

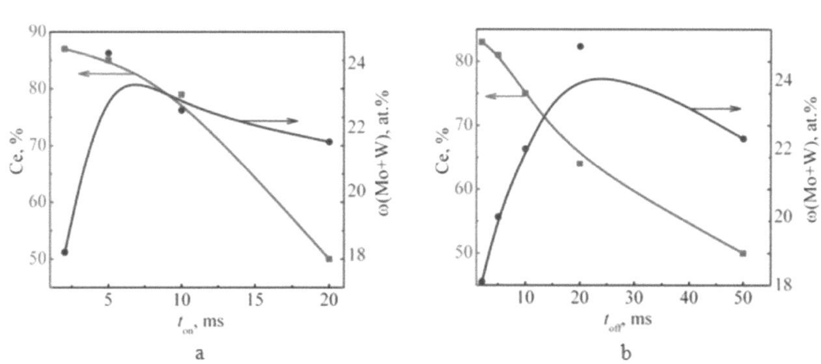

Figure 5.
Dependence of refractory metal content and current efficiency for Co—Mo—W alloys deposited from bi-ligand bath on parameters of pulse electrolysis: on-time (a) at t_{off} = *20 ms and off-time (b) at* t_{on} = *5 ms.*

the ratio of the molybdate and tungstate concentrations in a solution are the tools for ternary alloy composition control.

Current efficiency decreases in the range of 87–52% when increasing on-time is caused by the hydrogen evolution enhancement (**Figure 5a**). Prolong the pause as compared with pulse time positively influences the current efficiency as subsequent chemical reactions accompanying the alloying metals discharge are more fully. A larger current interruption than for 20 ms reduces the efficiency of the process (**Figure 5b**) as it was also observed in [34]. Thus, high refractory metal content in the Co—Mo—W alloy while maintaining a reasonable value of current efficiency *Ce* 70–75% at a current density of 10 A dm^{-2} is achieved with t_{on}/t_{off} = 5/20 ms.

Figure 6 shows X-ray diffraction patterns for Co—Mo—W alloys deposited on a copper substrate at a pulse current amplitude of 8 A dm^{-2}; t_{on}/t_{off} = 2/10 ms (**Figure 6**, black line) and a direct current density of 4 A dm^{-2} (**Figure 6**, red line); T = 30°C; coating thickness is 20 μm.

The X-ray diffraction patterns indicate an amorphous-crystalline structure of the alloys. The high intensity peaks at 60 and 90° are copper substrate lines. We can see some peaks corresponding to α-Co phase, intermetallic phase Co_7W_6 as previously for binary Co—W alloy [32], additional reflections of the intermetallic compound Co_7Mo_6, and rather wide halo with width about 15° is detected at angles 2θ of 43–58° (**Figure 6**) that reflects amorphous structure for both coatings deposited at pulse and direct current. The most important fact is the appearance of reflexes of metallic molybdenum and tungsten on XRD patterns for Co—Mo—W alloys deposited at pulse current. Such a character of X-ray diffraction patterns for coatings obtained by pulse current confirms our proposed mechanism of alloy formation [14, 28]. Metallic tungsten and molybdenum are formed in the chemical stage of reduction of refractory metal intermediate oxides during the pause of polarization. Such a pattern, along with decreasing tungsten content in the ternary alloy Co—Mo—W compared with a binary Co—W [32], also indicates molybdenum and tungsten competition during deposition in the alloy. The coherent-scattering region size of the amorphous part is 2–8 nm.

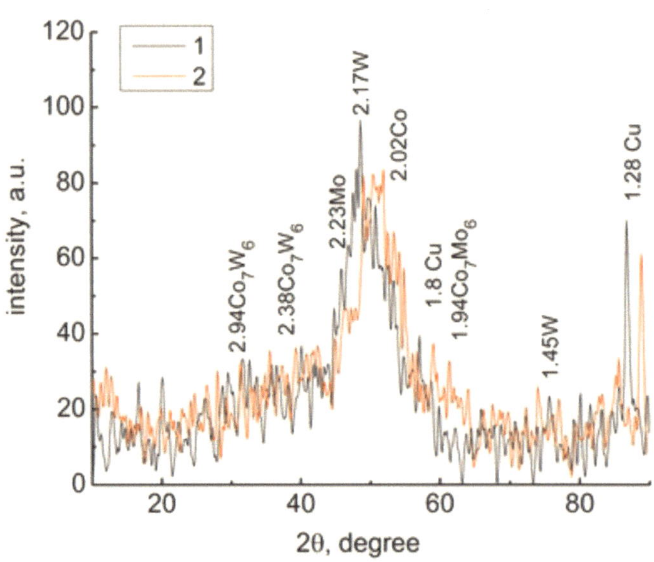

Figure 6.
X-ray diffraction patterns for electrolytic alloys Co—Mo—W of composition, at. %: (1) Co—75, Mo—16, W—9; (2) Co—85, Mo—9, W—5.

3. Co—Mo—Zr electrolytic alloys

The molybdenum content in Co—Mo—Zr alloys changes with current density increasing in the range of 2–4 A dm^{-2} (**Figure 7a**) similar to the Co—Mo—W films. At the same time, we observe a wider range of current densities of 4–8 A dm^{-2} providing coatings with molybdenum content of 24–25 at. %. Coating enrichment by molybdenum with increasing current density is entirely predictable since the molybdate electrochemical behavior is associated with multi-stage process following by chemical reducing of intermediate molybdenum oxides with hydrogen ad-atoms H_{ad} [27, 29, 35]. The cathode potential at Co—Mo—Zr electrodeposition is rather negative −(2.0–2.8) V (**Figure 7b**) and it becomes more negative with current density, that resulting in acceleration of side reaction producing H_{ad} which are involved in a chemical step of reducing intermediate molybdenum oxides. Exactly due to above reasons, the molybdenum content in the deposits is increased. However, at current densities above 8 A dm^{-2}, hydrogen evolution reaction becomes the dominant as evidenced by the decreasing current efficiency (**Figure 7b**), whereby the molybdenum content in the alloy decreases.

The zirconium content in the ternary coatings reaches 3.6–3.7 at. % with increasing current density up to 4 A dm^{-2} (**Figure 7a**). As it follows from the experimental data as well as previous investigations [26], molybdenum and zirconium competition is observed when codeposited to get the ternary alloy at higher current density than of 4 A dm^{-2}. Such behavior is associated with the different mechanisms of alloying metals reducing from citrate-pyrophosphate electrolyte. Really, molybdate reducing to metallic state proceeds through the six electrons transfer accompanied by the removal of four oxygen atoms. Therefore, deposits may also contain incompletely reduced intermediate molybdenum oxides. Zirconium is likely included in the deposit in the form of oxygen compounds ZrO_x due to higher binding energy Zr–O [31]. The EDS analysis data confirm the oxygen availability in the composition of the surface layers [31].

Nonlinear dependences of current efficiency *Ce* on the deposition current density (**Figure 7b**) were obtained, and *Ce* increases by 20% and reaches 63% with a rising current density from 5 to 8 A dm^{-2}; however, further increase in *i* reduces the current efficiency up to 47%. Such behavior may be attributed with acceleration of site hydrogen evolution reaction at more negative potentials.

Pulsed electrolysis allows the use of higher current densities, and not only energy parameter but also polarization time t_{on} and current interruption time t_{off} as well as its ratio are effectively used to control deposit composition and current efficiency. The shortest pulse duration should ensure the achievement of the alloy

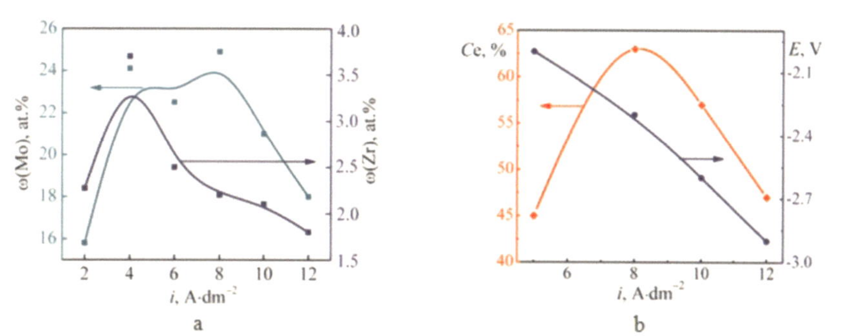

Figure 7.
Pulse current density influence on the composition (a) and current efficiency (b) for Co—Mo—Zr coatings; t_{on}/t_{off} *2/10 ms; T 20–25°C; pH 8; plated time 30 min.*

deposition potential, as well as t_{on} is limited by the requirements for the visual quality of coatings and efficiency of electrolysis while minimizing side reactions.

Electrodeposition of Co—Mo—Zr alloys is the relevant example to demonstrate the benefits and flexibility of pulsed electrolysis control. Molybdenum and zirconium content in the alloys deposited at a current density of 4 A dm^{-2} rises when increasing pulse time of 0.5–2 ms while maintaining pause time 10 ms (**Figure 8a**). Obviously, the real current value increases at the expense of a full signal handling, thereby achieving potential of alloying metal reduction in alloy. However, an increase in the pulse duration of more than 2 ms reduces the zirconium content; therefore, it does not seem appropriate.

As follows from the experimental data (**Figure 8b**) prolong current interruption time of 5–10 ms everything else being equal promotes zirconium content in the alloy of 2.1 up to 3.7 at. %; although increasing the pause reduces the incorporation of this metal in the alloy. Thus, the top zirconium content in the electrolytic deposits is reached at the ratio t_{on}/t_{off} = 2/10 ms (duty factor q = 10 and f = 85 Hz). It should be stated, the molybdenum percentage in Co—Mo—Zr deposits rises from 16.0 to 24.0 at. % with the pause duration (**Figure 8b**) due to more complete chemical reducing of intermediate molybdenum oxides by H_{ad}. It also confirms difference in zirconium and molybdenum reducing mechanism as well as their competition when deposited into the alloy.

As for Co—Mo—W alloy, current efficiency of Co—Mo—Zr deposition decreases with increasing pulse time due to acceleration of side reaction. Prolonging the pause contributes to the current efficiency as the following chemical reactions accompanying the alloying metal discharge are more full; and a larger current interruption reduces the efficiency of the process. Thus, current efficiency reaches maximum 98% when t_{off} = 50 ms and t_{on} = 2 ms.

Surface morphology of Co—Mo—Zr coatings changes with increasing the current density amplified internal stress that leads to fracture grid (**Figure 9**). The coating surface becomes less smooth and more globular, and the crystallite sizes increase at the higher polarization and respectively at larger molybdenum content (**Figure 9a** and **c**). EDS analysis data (**Figure 10**) show sufficiently uniform distribution of the alloying metals on uneven relief of the deposits which is typical for pulsed electrolysis and emphasizes its advantage over stationary. Increasing in polarization time at a fixed pause contributes growth of irregular spheroids on the surface; and microcracks become larger as observed in [36]. Furthermore, the coatings deposited at the current density of 6–8 A dm^{-2} and on-time of 10 ms are more porous as compared with other (**Figure 9b** and **c**), apparently due to accelerated hydrogen evolution.

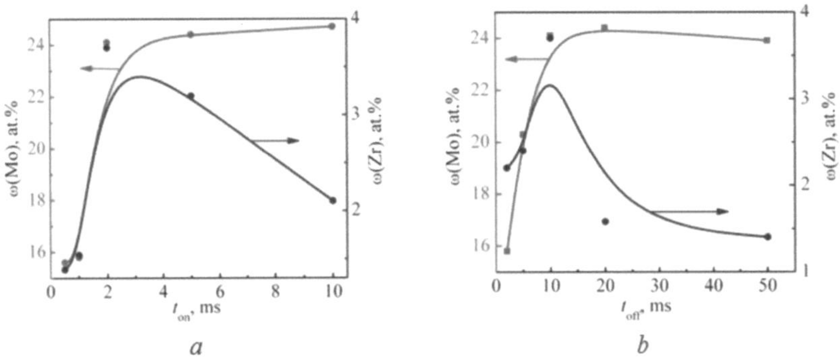

Figure 8.
Dependence of Co—Mo—Zr coating composition on the time of pulse t_{on} (a) (t_{off} 10 ms) and pause t_{off} (b) (t_{on} 2 ms); i = 4 A dm^{-2}; T = 20–25°C; pH 8; plated time 30 min.

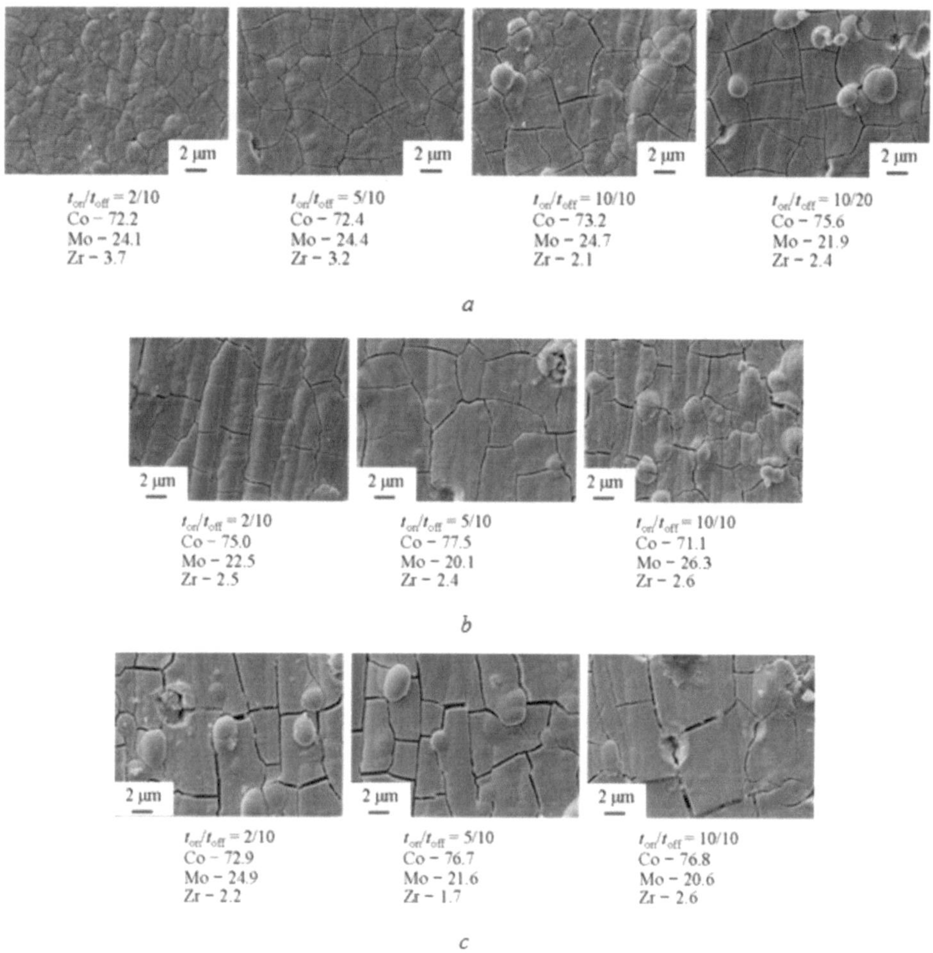

Figure 9.
Morphology (×2000) and composition (at. %) of Co—Mo—Zr coatings deposited in pulse mode at current density, A dm^{-2}*: 4 (a); 6 (b) and 8 (c); T = 20–25°C; pH 8; plated time 30 min.*

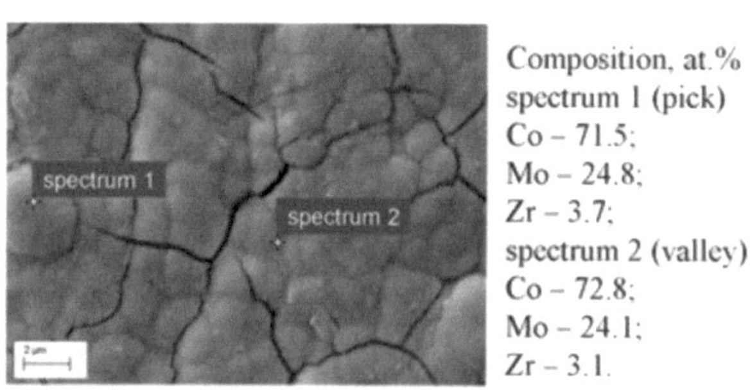

Figure 10.
Distribution of alloying elements at picks and valleys of coating Co—Mo—Zr deposited at 4 A dm^{-2}*,* t_{on}/t_{off} = 2/10.

Figure 11 shows X-ray diffraction patterns for electrolytic alloys Co—Mo—Zr deposited at a pulse current amplitude of 4 A dm^{-2} (**Figure 11** black line) and 8 A dm^{-2} (**Figure 11** red line). A series of diffraction lines for α-Co on X-ray diffraction patterns for Co—Mo—Zr deposits on steel substrates was obtained (**Figure 11**). The high intensity peaks at 52, 60, and 90° are copper substrate lines.

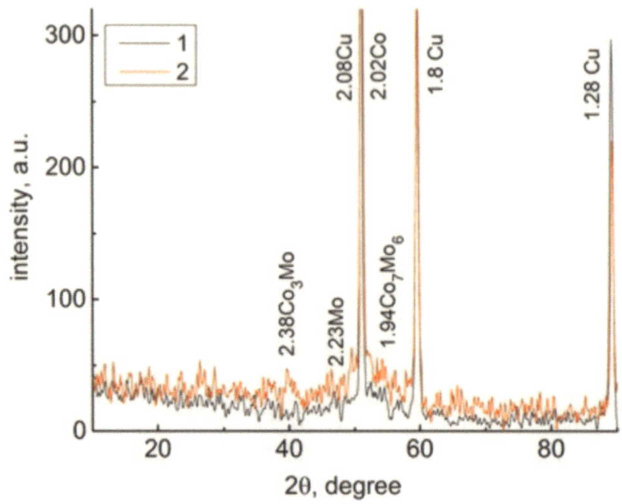

Figure 11.
X-ray diffraction patterns for electrolytic alloys Co—Mo—Zr of composition, at. %: (1) Co—72.2, Mo—24.1, Zr—3.7; (2) 72.9, Mo—24.9, Zr—2.2.

We can see peaks corresponding to intermetallic compounds Co_3Mo and Co_7Mo_6. Furthermore, one can find a small halo with full width at half maximum about 10° at angles 2θ 48–58° ~59°, which indicates an XRD amorphous structure of above materials [34]. Thus, the X-ray diffraction patterns indicate an amorphous-crystalline structure of the alloys. The most important fact is the appearance of reflexes of metallic molybdenum on XRD patterns for Co—Mo—Zr alloys deposited at higher current density 8 A dm^{-2} (**Figure 11** red line). In addition, the higher intensity of intermetallic compound reflexes is due to the enrichment of the alloy with the refractory component. The coherent-scattering region size of the amorphous part is 2–6 nm.

4. Corrosion behavior of binary and ternary cobalt electrolytic alloys

It is obvious that hydrogen ions H^+ are the oxidizing agents for cobalt-based electrolytic coatings in acidic medium, as well as in neutral and alkaline environment, only oxygen is a strong oxidizing agent [37]. Open circuit (corrosion) potentials E_{oc} of coated samples in corrosive solution with pH 3 are more positive than the potential of steel substrate which indicates the anodic control of corrosion process. The deep corrosion index k_h is reduced by almost three orders of magnitude (**Table 2**) and such inhibition in alloy corrosion is due to the acidic nature of refractory metals and zirconia oxides forming on the surface in oxygen containing media. The corrosion index of ternary alloys in acidic and neutral chloride-containing environment is almost halved compared with binary systems (**Table 2**). Such corrosion performance is associated with a decrease in roughness and smoothing out the relief of ternary coatings as compared with binary. In addition, the combined presence of molybdenum and tungsten or zirconium in the coatings provides a synergistic effect due to molybdenum and tungsten enhancing corrosion resistance to pitting as well as zirconium increasing tend to passivity. Moreover, with increasing total content of alloying metals, corrosion resistance increases.

Open circuit potentials of coated samples in corrosive solution with pH 11 are slow negative as compared with steel substrate (**Table 2**), that stipulates cathodic

Electrolytic alloy composition, at. %	Corrosive medium based on 1 M Na_2SO_4					
	pH 3		3% NaCl (pH 7)		pH 11	
	E_{oc}, V	k_h, mm/year	E_{oc}, V	k_h, mm/year	E_{oc}, V	k_h, mm/year
Steel substrate	−0.62	1.85	−0.56	0.93	−0.43	0.12
Co_{76}—Mo_{24}	−0.42	0.023	−0.25	0.029	−0.43	0.003
Co_{84}—W_{16}	−0.32	0.024	−0.35	0.01	−0.44	0.004
Co_{83}—Mo_{12}—W_5	−0.22	0.012	−0.37	0.007	−0.47	0.008
Co_{79}—Mo_{16}—W_5	−0.25	0.012	−0.38	0.005	−0.50	0.007
Co_{72}—Mo_{24}—Zr_4	−0.46	0.003	−0.5	0.002	−0.47	0.003
Co_{73}—Mo_{25}—Zr_2	−0.47	0.004	−0.5	0.003	−0.48	0.005

Table 2.
Corrosion indicators (E_{oc}, V; k_h, mm per year) of testing materials in different media.

Electrolytic alloy composition, at. %		Functional properties	Application area	Reference
Co Mo W	45–80 14–34 5–15	Hv 450–1100 MN/m^2 lgi^0_H ($Co_{71}Mo_{16}W_{13}$) = −3.35 [A/cm^2]	High hard, wear, heat resistant and corrosion protective coatings	[12, 15, 32]
Co Mo Zr	70–80 15–25 2–4	Hv 450–900 MN/m^2 lgi^0H ($Co_{75}Mo_{21}Zr_4$) = −3.1 [A cm^2]	Corrosion protective coatings, electrocatalytic films	[26, 31]

Table 3.
Application areas for the cobalt ternary coatings.

control of corrosion process. Such corrosion performance is mainly due to the basic nature of cobalt (II) oxides and hydroxides, which hinders the oxygen transport. The difference in corrosion rate of substrate and coated samples in such conditions is somewhat less—the deep corrosion index k_h is reduced by almost two orders of magnitude.

Summarizing and relying on the results of earlier studies [12, 15, 26, 31, 32] of the ternary cobalt alloy functional properties, it is possible to present their application areas in the table form (**Table 3**). This table reflects the composition, appropriative properties (microhardness Hv, corrosion resistance, and catalytic activity in hydrogen evolution reaction indicated as hydrogen exchange current density lgi^0_H), and technological application of designed coatings.

5. Summarizing the observations

Ternary Co—Mo—W electrolytic alloys deposited from citrate-pyrophosphate bath at pulse current composition and surface morphology were shown to be dependent on the current density and on/off time. The top refractory metal content in deposits was obtained at the current density of 9–10 A dm^{-2} and on/off time of 5/20 ms, but increasing current density diminishes efficiency of electrolysis. Tungsten content in the alloy was found to be much lower than molybdenum: W—2–7 at. % *vs*

Mo—16–22 at. % when depositing Co—Mo—W from the electrolyte with the oxometalate ratio Mo:W as 1:2. Structure of the ternary alloys was found to be amorphous-crystalline, and coherent-scattering region size was detected of 2–8 nm. Ternary coatings contain intermetallic phases Co_7W_6 and Co_7Mo_3. The amorphous structure of ternary alloys and significant content of alloying elements (Mo and W) provide high corrosion resistance increasing with total content of refractory metals.

Ternary galvanic alloys Co—Mo—Zr of different compositions and morphologies are deposited from citrate-pyrophosphate electrolyte at pulse current. Coatings with micro-globular morphology possessing low level of stress and cracks are formed at the current density of 4–6 A dm^{-2} and on/off time 2/10 ms. Structure of Co—Mo—Zr alloys was found to be amorphous crystalline, and coherent-scattering region size was detected of 2–6 nm. Ternary coatings contain intermetallic phases Co_3Mo and Co_7Mo_6. Co—Mo—Zr coatings high corrosion resistance is due to the molybdenum and zirconium tend to passivity, as well as the acid nature of their oxides.

The designed coating application areas corresponding to the functional properties are presented.

Author details

Gulmira Yar-Mukhamedova[1]*, Maryna Ved'[2], Nikolay Sakhnenko[2] and Tetiana Nenastina[2]

1 Institute of Experimental and Theoretical Physics of Al-Farabi Kazakh National University, Almaty, Kazakhstan

2 National Technical University "Kharkiv Polytechnic Institute", Kharkiv, Ukraine

*Address all correspondence to: gulmira-alma-ata@mail.ru

References

[1] Alexandrov Y, Pospelow A, Ved' M, et al. New application of electrode impedance spectroscopy in electrosynthesis of the organic conductors. In: Abstracts of NATO ARW "Recent Advances towards New Technologies"; 14-18 September 2003; Sudak, Crimea, Ukraine. 2003. p. 52

[2] Alexandrov Y, Sakhnenko N, Pospelow A, et al. Ion-selective electrodes based on organic conductors. In: Abstracts of the 5-th Int. Conf. "Electronic Processes in Organic Materials" (ICEPOM-5); 24-29 May 2004; Kyiv: Naukoviy Svit. 2004. pp. 209-210

[3] Sakhnenko MD, Ved' MV, Bairachna TM. Electrolytic alloys: Origin of synergetic effects. In: Materials of Conference Applied physico-inorganic chemistry; Simferopol: DIP. 2013. p. 53

[4] Yar-Mukhamedova GS. Influence of thermal treatment on corrosion resistance of chromium and nickel composite coatings. Materials Science. 2000;**36**:922-924. DOI: 10.1023/A:1011355325698

[5] Ved' M, Sakhnenko N, Bairachnaya T, Tkachenko N. Structure and properties of electrolytic cobalt-tungsten alloy coatings. Functional Materials. 2008;**15**:613-617

[6] Ahmad J, Asami K, Takeuchi A, Louzguine DV, Inoue A. High strength Ni-Fe-W and Ni-Fe-W-P alloys produced by electrodeposition. Materials Transactions. 2003;**44**(10): 1942-1947. Special issue on nano-hetero structures in advanced metallic materials

[7] Ved' MV, Sakhnenko ND, Karakurkchi AV, Zyubanova SI. Electrodeposition of iron-molybdenum coatings from citrate electrolyte. Russian Journal of Applied Chemistry. 2014;**87**:276-282. DOI: 10.1134/S1070427214030057

[8] Yao S, Zhao S, Guo H, Kowaka M. A new amorphous alloy deposit with high corrosion resistance. Corrosion. 1996; **52**(3):183-186. DOI: 10.5006/1.3292112

[9] Tsyntsaru N, Cesiulis H, Donten M, Sort J, Pellicer E, Podlaha-Murphy EJ. Modern trends in tungsten alloys electrodeposition with iron group metals. Surface Engineering and Applied Electrochemistry. 2012;**48**: 491-520. DOI: 10.3103/S1068375512060038

[10] Kohli S, Miglani S, Rapariya R. Basics of artificial neural network. International Journal of Computer Science and Mobile Computing. 2014;**3**: 745-751

[11] Ved' MV, Sakhnenko MD, Shtefan VV, et al. Computer modeling of the nonchromate treatment of aluminum alloys by neural networks. Materials Science. 2008;**44**(2):216-221

[12] Sakhnenko MD, Ved' MV, Ermolenko IY, Hapon YK, Kozyar MO. Design, synthesis, and diagnostics of functional galvanic coatings made of multicomponent alloys. Materials Science. 2017;**53**:680-686. DOI: 10.1007/s11003-017-0009-7

[13] Yermolenko IY, Ved' MV, Sakhnenko ND, Sachanova YI. Composition, morphology, and topography of galvanic coatings Fe-Co-W and Fe-Co-Mo. Nanoscale Research Letters. 2017;**12**:352. DOI: 10.1186/s11671-017-2128-3

[14] Karakurkchi AV, Ved' MV, Yermolenko IY, Sakhnenko ND. Electrochemical deposition of Fe–Mo–W alloy coatings from citrate electrolyte. Surface Engineering and

Applied Electrochemistry. 2016;**52**: 43-49. DOI: 10.3103/S1068375516010087

[15] Sakhnenko ND, Ved' MV, Hapon YK, Nenastina TA. Functional coatings of ternary alloys of cobalt with refractory metals. Russian Journal of Applied Chemistry. 2015;**87**:1941-1945. DOI: 10.1134/S107042721501200 6X

[16] Silkin SA, Gotelyak AV, Tsyntsaru N, Dikusar AI, Kreivaitis R, Padgurskas J. Effect of bulk current density on tribological properties of Fe-W, Co-W and Ni-W coatings. In: Proceedings of the 8th International Scientific Conference "BALTTRIB 2015". 2016. pp. 51-56. DOI: 10.15544/balttrib.2015.10

[17] Glushkova MA, Bairachna TN, Ved' MV, Sakhnenko ND. Electrodeposited cobalt alloys as materials for energy technology. In: Mater. Res. Soc. Symp. Proc.; USA, Boston. Vol. 1491. 2013. p. mrsf12-1491-c08-15. DOI: 10.1557/opl.2012.1672

[18] Ma SL, Xi X, Nie Z, Dong T, Mao Y. Electrodeposition and characterization of Co-W alloy from regenerated tungsten. International Journal of Electrochemical Science. 2017;**12**: 1034-1051. DOI: 10.20964/2017.02.37

[19] Capel H, Shipway PH, Harris SJ. Sliding wear behaviour of electrodeposited cobalt–tungsten and cobalt–tungsten–iron alloys. Wear. 2003;**255**:917-923. DOI: 10.1016/S0043-1648(03)00241-2

[20] Gomez E, Pellicer E, Alcobe X, Valles E. Properties of Co-Mo coatings obtained by electrodeposition at p. 6.6. Journal of Solid State Electrochemistry. 2004;**8**:497-504. DOI: 10.1007/s10008-004-0495-z

[21] Gomez E, Pellicer E, Valles E. Electrodeposited cobalt_molybdenum magnetic materials. Journal of Electroanalytical Chemistry. 2001;**517**: 109-116. DOI: 10.1016/S0022-0728(01)00682-9

[22] Gomez E, Kipervaser ZG, Pellicer E, Valles E. Extracting deposition parameters for cobalt–molybdenum alloy from potentiostatic current transients. Physical Chemistry Chemical Physics. 2004;**6**:1340-1344. DOI: 10.1039/B315057G

[23] Yar-Mukhamedova G, Ved' M, Sakhnenko N, Karakurkchi A, Yermolenko I. Iron binary and ternary coatings with molybdenum and tungsten. Applied Surface Science. 2016; **383**:346-352. DOI: 10.1016/j.apsusc.2016.04.046

[24] Karakurkchi AV, Ved' MV, Sakhnenko ND, Ermolenko IY. Electrodeposition of iron–molybdenum–tungsten coatings from citrate electrolytes. Russian Journal of Applied Chemistry. 2015;**88**:1860-1869. DOI: 10.1134/S107042721501018X

[25] Yermolenko IY, Ved' MV, Karakurkchi AV, Sakhnenko ND, Kolupayeva ZI. The electrochemical behavior of Fe^{3+}–WO_4^{2-}–Cit^{3-} and Fe^{3+}–MoO_4^{2-}–WO_4^{2-}–Cit^{3-} systems. Issues of Chemistry and Chemical Technology. 2017;**2**:4-14

[26] Yar-Mukhamedova G, Ved' M, Sakhnenko N, Koziar M. Ternary cobalt-molybdenum-zirconium coatings for alternative energies. Applied Surface Science. 2017;**421**(Part A):68-76. DOI: 10.1016/j.apsusc.2017.01.196

[27] Yar-Mukhamedova GS, Sakhnenko ND, Ved' MV, Yermolenko IY, Zyubanova SI. Surface analysis of Fe-Co-Mo electrolytic coatings. In: IOP Conference Series: Materials Science and Engineering. Vol. 213. 2017. DOI: 10.1088/1757-899X/213/1/012019

[28] Ved' MV, Ermolenko IY, Sakhnenko ND, Zyubanova SI, Sachanova YI. Methods for controlling the composition

and morphology of electrodeposited Fe–Mo and Fe–Co–Mo coatings. Surface Engineering and Applied Electrochemistry. 2017;**53**(6):525-532. DOI: 10.3103/S1068375517060138

[29] Shao II, Vereecken PM, Chien CL, Cammarata RC, Searson PC. Electrochemical deposition of FeCo and FeCoV alloys. Journal of the Electrochemical Society. 2003;**150**: C184-C188. DOI: 10.1149/1.1553789

[30] Ved' MV, Sakhnenko ND, Yermolenko IY, Nenastina TA. Nanostructured functional coatings of iron family metals with refractory elements. Chapter 1. In: Fesenko O, Yatsenko L, editors. Nanochemistry, Biotechnology, Nanomaterials, and Their Applications, Springer Proceedings in Physics. Vol. 214. Cham, Switzerland: Springer International Publishing AG, part of Springer Nature; 2018. pp. 3-34. DOI: 10.1007/978-3-319-92567-7_1

[31] Ved' M, Sakhnenko N, Koziar M, Slavkova M. Functional properties of electrolytic alloys of cobalt with molybdenum and zirconium. Functional Materials. 2016;**23**:420-426. DOI: 10.15407/fm23.03.420

[32] Yar-Mukhamedova G, Ved' M, Sakhnenko N, Nenastina T. Electrodeposition and properties of binary and ternary cobalt alloys with molybdenum and tungsten. Applied Surface Science. 2018;**445**:298-307. DOI: 10.1016/j.apsusc.2018.03.171

[33] Schlesinger M, Paunovic M. Modern Electroplating. 5th ed. Hoboken, NJ: John Wiley & Sons; 2010

[34] Ramanauskas R, Gudavičiūtė L, Juškėnas R. Effect of pulse plating on the composition and corrosion properties of Zn–Co and Zn–Fe alloy coatings. Chemija. 2008;**19**:7-13

[35] Tsyntsaru N, Cesiulis H, Budreika A. The effect of electrodeposition conditions and post-annealing on nanostructure of Co–W coatings. Surface and Coating Technology. 2012; **206**:4262-4269. DOI: 10.1016/j.surfcoat.2012.04.036

[36] Pustovalov EV, Modin EB, Voitenko OV, Fedorets AN, Dubinets AV, Grudin BN, et al. Structure relaxation and crystallization of the CoW-CoNiW-NiW electrodeposited alloys. Nanoscale Research Letters. 2014;**9**:66. DOI: 10.1186/1556-276X-9-66

[37] Tsyntsaru N, Dikusar A, Cesiulis H, Celis J-P, Bobanova Z, Sidel'nikova S, et al. Tribological and corrosive characteristics of electrochemical coatings based on cobalt and iron superalloys. Powder Metallurgy and Metal Ceramics. 2009;**48**:419-428. DOI: 10.1007/s11106-009-9150-7

Chapter 6

Optimization of Surface Roughness of D2 Steels in WEDM using ANN Technique

Umesh K. Vates, N.K. Singh, B.P. Sharma and S. Sivarao

Abstract

Attempt has been made to investigate the experimental process and surface roughness (SR) optimization of cold working (high carbon and high chromium) hard die steel (D2) during wire electrical discharge machining processes (WEDM). It is very difficult to determine optimal cutting parameters for improving cutting performance has been reported. Wire electrical discharge machining process relies heavily on the operators' technologies and experience because of their numerous and diverse range as using complicated cuts can made through difficult to machine electrically conductive components, WEDM process was developed to generate precise cutting on complicate, hard and difficult to machine materials. Tan-sigmoid and purlin transfer functional with bias based four layered back propagation artificial neural network (BPANN) approach have been used to investigate the effect of six independent parameters namely gap voltage (V_g), flush rate (F_r), Pulse on time (T_{on}), pulse off time (T_{off}), wire feed (W_f) and wire tension (W_t) over CLA value of surface roughness (R_a) along with corresponding material removal rate (MRR). A fractional factorial design of experiment of three level were employed to conduct 80 rows of experiment on (D2) steel with chrome coated copper alloy wire electrode. The predicted response, CLA values of SR and corresponding MRR were observed by the approach of BPANN from experimental (55 rows for training, 15 rows for validation and 10 for testing) data. Software instructed programme has been used individually for training, validation and testing in MATLAB 2010a to find the corresponding prediction output. Two fold cross over technique (TFCT) were used to developed distinguish (S1 and S2) models and also developed more models depending on numbers of neurons used in primary and secondary hidden layers. The model adequacy is very satisfactory as correlation coefficient (R^2) is found to be 99.1% and adjusted ($R_{adj.}^2$) statistics is 98.5. It is found those spark time ON/OFF, wire feed rate, wire tension, gap voltage and flush rate and few of their interactions have significant effect on SR.

Keywords: WEDM, BPANN, SR, MRR, TFCT

1. Introduction

Wire electrical discharge machining is the metal removal process by means of repeated spark created between the wire electrode and work piece. It is considered as unique adaptation of the conventional EDM, which used an electrode to create the sparking within kerfs [11]. However, WEDM utilizes a continuously traveling chromium coated copper wire electrode ranging diameter 0.05–0.35 mm, which is capable to achieve very good sharpness of edge [4]. Very high temperature ranging 8000–10,000°C creates within the kerfs gap during machining, so that material removal may takes place by not only melting but directly vaporizations also. WEDM is used for the high precision machining to all type of electrically conductive metallic alloys, tool and die, graphite, and few ceramic and composite materials of any hardness which cannot be machined easily by conventional machining methods [1, 5].

Manufacturing processes (WEDM) has been chosen depending on the material characteristics and the type of responses required to be evaluating. The present study aimed to optimization of responses i.e. surface roughness with corresponding MRR of D2 steel by conducting 80 rows of experimental data using frictional factorial (2^{6-2}) design of experiment of five different set at three levels [3]. Four layered BPANN architecture has been used for modeling, where independent process variables are V_g, F_r T_{on}, T_{off}, W_f and W_t to get the précised and optimized values of responses R_a [6, 8, 10]. Best model S2 has been found on the basis of correlation coefficient (R^2) between observed and predicted responses (SR) [12]. The response (SR) is expressed as the irregularities of material resulted from various machining operations. It is represented as '$\mathbf{R_a}$' symbol and used to be called center line arithmetic average roughness for the sampling length [2].

The optimum process parameters are much essential to achieve better surface finish with adequate material removal rate (MRR) or shrink of total machining time; lot of research attempts has been reported for modeling and investigation of WEDM process parameters [7], but sum of root mean square error (SRMSE) approach have been used to optimize the process parameters by taking 55 rows of training data [9].

2. Experimental setup

2.1 Selection of wire electrode and work piece

A chrome coated cylindrical pure copper wire electrode having 0.25 mm in diameter and high tensile strength were selected for conducting machining operation on 18 mm diameter of D2 steel rod to cut 5 mm thickness of disk using Electronica Maxicut, WEDM process. It is very clear that D2 is hard die steel and conducting material with high carbon and chromium content (**Table 1**).

The experiment has carried out on Wire Electrical Discharge Machine, model ELECTRONICA-MAXICUT, SLNO -250, (F:09:0002:01) having the facilities to hold the work piece within the place provided by the help of conductive fixture, so that they can complete the circuit between electrode and work piece. The spark is created depending upon gap voltage applied between the conductive work piece, electrode, and machining performance influence the major independent process parameter which selected for experiment as characteristics of screening test.

C	SI	Cr	Mo	V	HRC	Conductivity
1.50%	0.30%	12.00%	0.80%	0.90%	56	22 (W/mk)

Table 1.
Metallurgical component analysis: D2 steel.

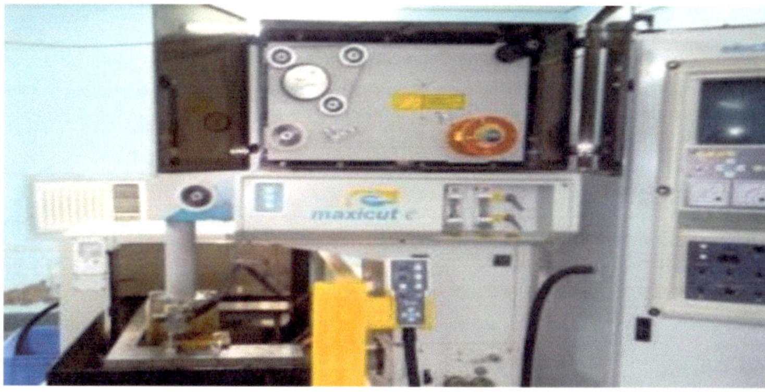

Figure 1.
D2 steel machining using WEDM process.

Figure 2.
Surftest SJ-210 (Mitutoyo).

Commercials grade of deionized water (density = 832 kg/m^3) was used as dielectric fluid. 18 mm cylindrical rod of D2 steel was used as the work piece with negative polarity and the power supply has the provision to connect the 0.25 mm chromium coated pure copper tool electrode with positive polarity so that the material removal may takes place by influence of heat generated within kerfs due to applied voltage within it (**Figurc 1**).

The surface roughness R_a of the processed material have been measured precisely by using Surftest SJ-210 tester having center line average value (CLA), where least count of the equipment is 0.001 μm for the travel length of 0.85 mm (**Figure 2**).

2.2 Design of experiment and objective

Five different set of fractional factorial (2^{6-2} = 16) experimental design have been selected at two levels, so that 80 rows of experimental data can be observed at three level of replication on D2 using WEDM. In this study the main aim to minimize the surface roughness of D2 on best possible maximum MRR during WEDM (**Table 2**).

2.3 ANN architecture and training

The hit and trail method based on literature have been adapted to find 7 and 10 neurons in primary and secondary hidden layers respectively, which effects on the R-square statistics for best prediction modeling. Tan sigmoid activation (squashing) function used as the (infinite input to finite output range) learning capability by the

Factors/three level (coding)	1	2	3
Gap voltage (Vg): (volt)	30	60	90
Flush rate (Fr): (L/min)	4	6	8
Pulse on time (Ton): (μS)	1.05	1.15	1.25
Pulse of time (Toff): (μS)	130	160	190
Wire feed rate (Wf):(m/min)	2	5	8
Wire tension (Wt): (g)	300	600	900

Table 2.
Factors for screening test.

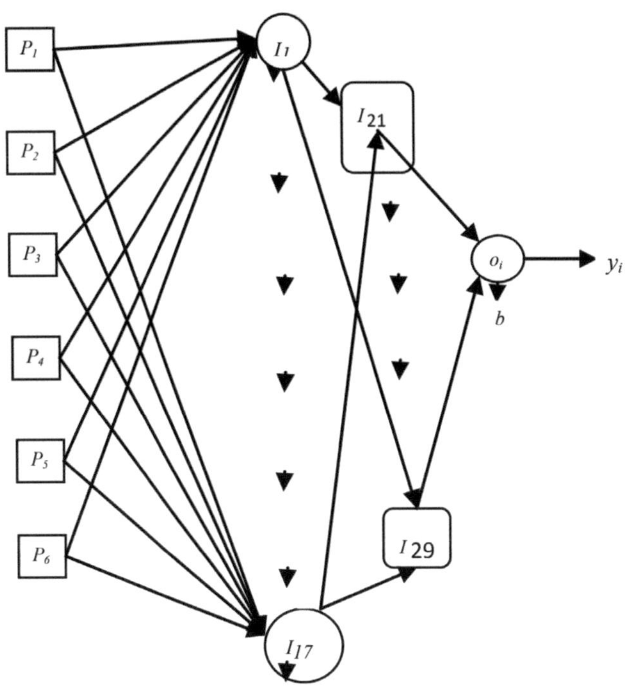

Figure 3.
Artificial neural network approach.

controllable instructed programme in MATLAB 2010a. Steepest descent problem used for the training algorithm to train the multilayer network, where the values of gradient was smallest because of the small changes in weight and biases. p_1, p_2, p_3, p_4, p_5 and p_6 are the six input layer neurons and O_i is the single neurons in output layer, whereas I_{11}-I_{17} and I_{21}-I_{29} (7 neurons present in primary and 10 in secondary hidden layers) are the hidden layers (**Figure 3**).

3. Experimentation

Two models (S1 and S2) were developed from 80 rows of experimental data performed of D2. Only training result of best performing model (S2) of 55 rows is

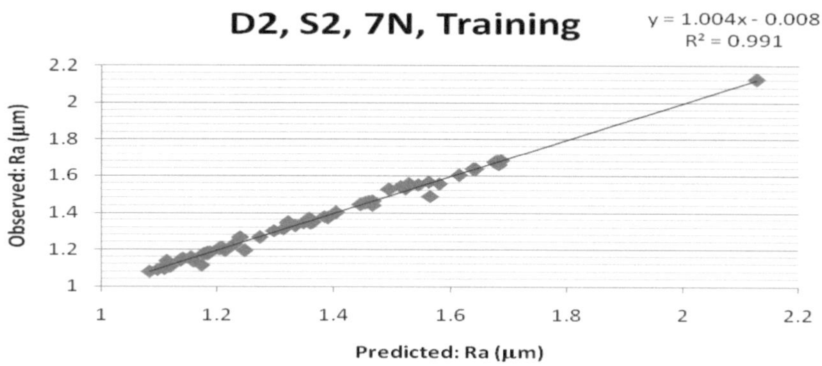

Figure 4.
Predictions against observations of R_a for model-D2, S2, 7 N (training dataset).

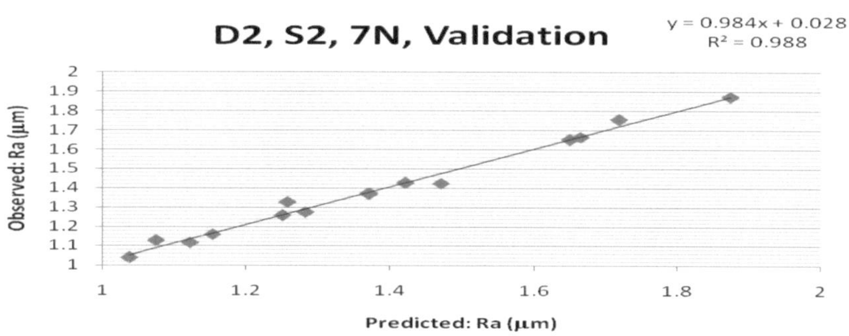

Figure 5.
Predictions against observations of R_a for model-D2, S2, 7 N (validation dataset).

presented here for achieving the aimed to optimization of influencing process parameters (**Figures 4** and **5**).

OPTIMIZATION OF PROCESS PARAMETER R_a: D2, 7 Neurons in hidden layer.

The best model needs to be predicted among Model-S1 and S2, in D2 steel. Effect of individual input parameters will be observed on the R_a (**Tables 3–6**).

4. Optimization of process parameters

It is evident from **Table 3**, that each independent influencing input parameter has corresponding values of their square of residuals at each three levels. Two values at each level (2 × 3 = 6 rows) has been taken for each inputs, where lowest possible square of residuals are available, to draw the **Figure 6(a–f)**.

5. Result

Figure 6(a–f) shows the relations between individual influencing parameters (V_g, F_r, T_{on}, T_{off}, W_f and W_t) to their optimized response, surface roughness (R_a) with corresponding values of MRR. **Table 5** also indicates that unique values of each influencing parameters (corresponding to its serial numbers of **Table 5**) gives optimum responses, which has been highlighted.

SN	Gap voltage (V_g)	Flush rate (F_r)	Spark time (T_{ON})	Spark time (T_{OFF})	Wire feed (W_f)	Wire tension (W_t)	Surface roughness (R_a) Obs.	Surface roughness (R_a) Pred.	Material removal (MRR)	Square of residuals
	(V)	(Lit./min)	(μS)	(μS)	(m/min)	(g)	(μm)	(μm)	(mg/min)	$(\mu m)^2$
1	30	4	1.05	130	2	300	1.6858	1.6863	102	2.5E−07
2	30	4	1.05	160	2	600	1.4452	1.4451	92	1E−08
3	30	4	1.15	130	5	600	1.3884	1.3713	133	0.0002924
4	30	4	1.15	160	5	300	1.4658	1.4428	95	0.000529
5	30	6	1.05	130	5	600	1.3836	1.3788	125	2.304E−05
6	30	6	1.05	160	5	300	1.5278	1.5553	110	0.0007562
7	30	6	1.15	130	2	300	1.676	1.6756	97	1.6E−07
8	30	6	1.15	160	2	600	1.564	1.4909	95	0.0053436
9	60	4	1.05	130	5	300	1.1772	1.1754	104	3.24E−06
10	60	4	1.05	160	5	600	1.2076	1.2083	88	4.9E−07
11	60	4	1.15	130	2	600	1.273	1.2663	136	4.489E−05
12	60	4	1.15	160	2	300	1.3476	1.3455	116	4.41E−06
13	60	6	1.05	130	2	600	1.3322	1.3277	110	2.025E−05
14	60	6	1.05	160	2	300	1.1598	1.1371	115	0.0005153
15	60	6	1.15	130	5	300	1.248	1.1945	118	0.0028623
16	30	8	1.15	160	8	900	1.5124	1.5422	145	0.000888
17	30	8	1.15	190	8	600	1.363	1.3482	108	0.000219
18	30	8	1.25	160	5	600	2.1256	2.128	206	5.76E−06
19	30	8	1.25	190	5	900	1.6794	1.6823	101	8.41E−06
20	90	4	1.15	160	8	600	1.1098	1.1096	88	4E−08
21	90	4	1.15	190	8	900	1.1096	1.0952	63	0.0002074

SN	Gap voltage (V_g)	Flush rate (F_r)	Spark time (T_{ON})	Spark time (T_{OFF})	Wire feed (W_f)	Wire tension (W_t)	Surface roughness (R_a) Obs.	Surface roughness (R_a) Pred.	Material removal (MRR)	Square of residuals
	(V)	(Lit./min)	(μS)	(μS)	(m/min)	(g)	(μm)	(μm)	(mg/min)	$(\mu m)^2$
22	90	4	1.25	160	5	900	1.3572	1.3664	107	8.464E-05
23	90	4	1.25	190	5	600	1.3218	1.3425	88	0.0004285
24	90	8	1.15	160	5	900	1.2286	1.2292	91	3.6E−07
25	90	8	1.15	190	5	600	1.1194	1.1062	64	0.0001742
26	60	6	1.15	160	5	600	1.4038	1.4023	155	2.25E−06
27	60	8	1.05	130	5	900	1.4592	1.459	162	4E−08
28	60	8	1.05	160	5	600	1.3601	1.3441	139	0.000256
29	60	8	1.25	130	2	600	1.5208	1.5302	202	8.836E−05
30	60	8	1.25	160	2	900	1.5435	1.5535	168	0.0001
31	90	6	1.05	130	5	600	1.3127	1.3118	78	8.1E−07
32	90	6	1.05	160	5	900	1.2973	1.3023	72	2.5E−05
33	90	6	1.25	130	2	900	1.1823	1.1867	117	1.936E−05
34	90	6	1.25	160	2	600	1.0832	1.0812	105	4E−06
35	90	8	1.05	130	2	900	1.2396	1.2696	89	0.0009
36	90	8	1.05	160	2	600	1.1838	1.1739	81	9.801E−05
37	90	8	1.25	130	5	600	1.1413	1.1524	92	0.0001232
38	90	8	1.25	160	5	900	1.1125	1.1364	112	0.0005712
39	60	6	1.05	130	2	600	1.4536	1.4546	128	1E−06
40	60	6	1.05	160	2	900	1.3208	1.3474	114	0.0007076
41	90	8	1.05	130	2	900	1.1369	1.1423	96	2.916E−05
42	90	8	1.05	160	2	600	1.0962	1.0905	78	3.249E−05

SN	Gap voltage (V_g)	Flush rate (F_r)	Spark time (T_{ON})	Spark time (T_{OFF})	Wire feed (W_f)	Wire tension (W_t)	Surface roughness (R_a) Obs.	Surface roughness (R_a) Pred.	Material removal (MRR)	Square of residuals
	(V)	(Lit./min)	(μS)	(μS)	(m/min)	(g)	(μm)	(μm)	(mg/min)	$(\mu m)^2$
43	90	8	1.25	130	5	600	1.1551	1.1551	99	0
44	90	8	1.25	160	5	900	1.1723	1.1153	74	0.003249
45	30	4	1.15	160	2	300	1.6813	1.6628	112	0.0003422
46	30	4	1.15	190	2	900	1.5782	1.5577	108	0.0004202
47	30	4	1.25	160	8	900	1.4935	1.5283	163	0.001211
48	30	4	1.25	190	8	300	1.4658	1.4666	155	6.4E−07
49	30	6	1.15	160	8	900	1.6402	1.6368	121	1.156E−05
50	30	6	1.15	190	8	300	1.6128	1.6021	132	0.0001145
51	30	6	1.25	160	2	300	1.6368	1.6354	103	1.96E−06
52	30	6	1.25	190	2	900	1.5609	1.5668	108	3.481E−05
53	60	4	1.15	160	8	300	1.2136	1.1945	123	0.0003648
54	60	4	1.15	190	8	900	1.1871	1.1878	128	4.9E−07
55	60	4	1.25	160	2	900	1.2036	1.2035	148	1E−08
				Average				1.3654	113.8	

Table 3.
D2, S1, 7N, training data (combined parameters).

SN	Gap voltage (V_g)	Flush rate (F_r)	Spark ON time (T_{ON})	Spark OFF time (T_{OFF})	Wire feed (W_f)	Wire tension (W_t)	Surface roughness (R_a) obs.	Surface roughness (R_a) predicted.	$(\text{Residual})^2$	Material removal predicted (MRR)
	(V)	(Lit./min)	(μS)	(μS)	(m/min)	(g)	(μm)	(μm)	$(\mu m)^2$	(mg/min)
1	30	4	1.05	130	2	300	1.6858	1.6863	2.5E−07	105
2	30	4	1.05	160	2	600	1.4452	1.4451	1E−08	95
3	30	4	1.15	130	5	600	1.3884	1.3713	0.0002924	119
4	30	4	1.15	160	5	300	1.4658	1.4428	0.000529	102
5	30	6	1.05	130	5	600	1.3836	1.3788	2.304E−05	115
6	30	6	1.05	160	5	300	1.5278	1.5553	0.0007562	116
7	30	6	1.15	130	2	300	1.676	1.6756	1.6E−07	114
8	30	6	1.15	160	2	600	1.564	1.4909	0.0053436	102
9	60	4	1.05	130	5	300	1.1772	1.1754	3.24E−06	108
10	60	4	1.05	160	5	600	1.2076	1.2083	4.9E−07	96
11	60	4	1.15	130	2	600	1.273	1.2663	4.489E−05	131
12	60	4	1.15	160	2	300	1.3476	1.3455	4.41E−06	123
13	60	6	1.05	130	2	600	1.3322	1.3277	2.025E−05	111
14	60	6	1.05	160	2	300	1.1598	1.1371	0.0005153	117
15	60	6	1.15	130	5	300	1.248	1.1945	0.0028623	112
16	30	8	1.15	160	8	900	1.5124	1.5422	0.000888	136
17	30	8	1.15	190	8	600	1.363	1.3482	0.000219	105
18	30	8	1.25	160	5	600	2.1256	2.128	5.76E−06	189
19	30	8	1.25	190	5	900	1.6794	1.6823	8.41E−06	97
20	90	4	1.15	160	8	600	1.1098	1.1096	4E−08	79

SN	Gap voltage (V_g)	Flush rate (F_r)	Spark ON time (T_{ON})	Spark OFF time (T_{OFF})	Wire feed (W_f)	Wire tension (W_t)	Surface roughness (R_a) obs.	Surface roughness (R_a) predicted.	$(\text{Residual})^2$	Material removal predicted (MRR)
	(V)	(Lit./min)	(μS)	(μS)	(m/min)	(g)	(μm)	(μm)	$(\mu m)^2$	(mg/min)
21	90	4	1.15	190	8	900	1.1096	1.0952	0.0002074	70
22	90	4	1.25	160	5	900	1.3572	1.3664	8.464E−05	110
23	90	4	1.25	190	5	600	1.3218	1.3425	0.0004285	92
24	90	8	1.15	160	5	900	1.2286	1.2292	3.6E−07	101
25	90	8	1.15	190	5	600	1.1194	1.1062	0.0001742	69
26	60	6	1.15	160	5	600	1.4038	1.4023	2.25E−06	153
27	60	8	1.05	130	5	900	1.4592	1.459	4E−08	158
28	60	8	1.05	160	5	600	1.3601	1.3441	0.000256	143
29	60	8	1.25	130	2	600	1.5208	1.5302	8.836E−05	208
30	60	8	1.25	160	2	900	1.5435	1.5535	0.0001	163
31	90	6	1.05	130	5	600	1.3127	1.3118	8.1E−07	74
32	90	6	1.05	160	5	900	1.2973	1.3023	2.5E−05	93
33	90	6	1.25	130	2	900	1.1823	1.1867	1.936E−05	122
34	90	6	1.25	160	2	600	1.0832	1.0812	4E−06	111
35	90	8	1.05	130	2	900	1.2396	1.2696	0.0009	97
36	90	8	1.05	160	2	600	1.1838	1.1739	9.801E−05	86
37	90	8	1.25	130	5	600	1.1413	1.1524	0.0001232	81
38	90	8	1.25	160	5	900	1.1125	1.1364	0.0005712	106
39	60	6	1.05	130	2	600	1.4536	1.4546	1E−06	135
40	60	6	1.05	160	2	900	1.3208	1.3474	0.0007076	111

SN	Gap voltage (V_g)	Flush rate (F_r)	Spark ON time (T_{ON})	Spark OFF time (T_{OFF})	Wire feed (W_f)	Wire tension (W_t)	Surface roughness (R_a) obs.	Surface roughness (R_a) predicted.	$(Residual)^2$	Material removal predicted (MRR)
	(V)	(Lit./min)	(μS)	(μS)	(m/min)	(g)	(μm)	(μm)	$(\mu m)^2$	(mg/min)
41	90	8	1.05	130	2	900	1.1369	1.1423	2.916E−05	96
42	90	8	1.05	160	2	600	1.0962	1.0905	3.249E−05	74
43	90	8	1.25	130	5	600	1.1551	1.1551	0	94
44	90	8	1.25	160	5	900	1.1723	1.1153	0.003249	88
45	30	4	1.15	160	2	300	1.6813	1.6628	0.0003422	117
46	30	4	1.15	190	2	900	1.5782	1.5577	0.0004202	100
47	30	4	1.25	160	8	900	1.4935	1.5283	0.001211	158
48	30	4	1.25	190	8	300	1.4658	1.4666	6.4E−07	163
49	30	6	1.15	160	8	900	1.6402	1.6368	1.156E−05	115
50	30	6	1.15	190	8	300	1.6128	1.6021	0.0001145	141
51	30	6	1.25	160	2	300	1.6368	1.6354	1.96E−06	112
52	30	6	1.25	190	2	900	1.5609	1.5668	3.481E−05	109
53	60	4	1.15	160	8	300	1.2136	1.1945	0.0003648	123
54	60	4	1.15	190	8	900	1.1871	1.1878	4.9E−07	125
55	60	4	1.25	160	2	900	1.2036	1.2035	1E−08	144
				Average				1.3654	0.002642	114.8

Table 4.
Training data for model: S2, R_a, N7, D2 steel.

SN	Gap voltage (V_g)	Flush rate (F_r)	Spark time (T_{ON})	Spark time (T_{OFF})	Wire feed (W_f)	Wire tension (W_t)	Surface roughness (R_a)	Material removal (MRR)	Square of residuals (R_a)
	(V)	(Lit./min)	(µS)	(µS)	(m/min)	(g)	(µm)	(mg/min)	$(\mu m)^2$
	V_g								
2	30	4	1.05	160	2	600	1.4452	92	1E−08
7	30	6	1.15	130	2	300	1.676	97	1.6E−07
27	60	8	1.05	130	5	900	1.4592	162	4E−08
54	60	4	1.15	190	8	900	1.1871	128	4.9E−07
20	90	4	1.15	160	8	600	1.1098	88	4E−08
43	90	8	1.25	130	5	600	1.1551	99	0
		F_r							
1	30	4	1.05	130	2	300	1.6858	102	2.5E−07
55	60	4	1.25	160	2	900	1.2036	148	1E−08
7	30	6	1.15	130	2	300	1.676	97	1.6E−07
31	90	6	1.05	130	5	600	1.3127	78	8.1E−07
27	60	8	1.05	130	5	900	1.4592	162	4E−08
43	90	8	1.25	130	5	600	1.1551	99	0
			T_{on}						
41	90	8	1.05	130	2	900	1.1369	96	2.916E−05
42	90	8	1.05	160	2	600	1.0962	78	3.249E−05
54	60	4	1.15	190	8	900	1.1871	128	4.9E−07
20	90	4	1.15	160	8	600	1.1098	88	4E−08
55	60	4	1.25	160	2	900	1.2036	148	1E−08
43	90	8	1.25	130	5	600	1.1551	99	0
				T_{off}					
7	30	6	1.15	130	2	300	1.676	97	1.6E−07
27	60	8	1.05	130	5	900	1.4592	162	4E−08
55	60	4	1.25	160	2	900	1.2036	148	1E−08
36	90	8	1.05	160	2	600	1.1838	81	9.801E−05
21	90	4	1.15	190	8	900	1.1096	63	0.0002074
54	60	4	1.15	190	8	900	1.1871	128	4.9E−07
					W_f				
34	90	6	1.25	160	2	600	1.0832	105	4E−06
2	30	4	1.05	160	2	600	1.4452	92	1E−08
43	90	8	1.25	130	5	600	1.1551	99	0
31	90	6	1.05	130	5	600	1.3127	78	8.1E−07
48	30	4	1.25	190	8	300	1.4658	155	6.4E−07
16	30	8	1.15	160	8	900	1.5124	145	0.000888

SN	Gap voltage (V_g)	Flush rate (F_r)	Spark time (T_{ON})	Spark time (T_{OFF})	Wire feed (W_f)	Wire tension (W_t)	Surface roughness (R_a)	Material removal (MRR)	Square of residuals (R_a)
	(V)	(Lit./min)	(μS)	(μS)	(m/min)	(g)	(μm)	(mg/min)	$(\mu m)^2$
						W_t			
7	30	6	1.15	130	2	300	1.6760	97	1.6E−07
1	30	4	1.05	130	2	300	1.6858	102	2.5E−07
2	30	4	1.05	160	2	600	1.4452	92	1E−08
31	90	6	1.05	130	5	600	1.3127	78	8.1E−07
54	60	4	1.15	190	8	900	1.1871	128	4.9E−07
21	90	4	1.15	190	8	900	1.1096	63	0.0002074

Correlation coefficient (R^2): Training data of D2 steel (best performing model S2) using 7 and 10 neurons in primary and secondary hidden layers.

Table 5.
D2, S2, 7N, training data (individual parameters corresponding to least square of residuals).

Material	Model	R^2 value	Equation of lines (correlation between obs. and pred. values of Ra)	Average predicted R_a (μm)	Root mean square error (μm)	Percentage RMSE (%)	Average % RMSE RMSE
D2	S1 Training	0.983	y = 1.005x - 0.010	1.3864	0.003401	0.2453	0.8129 **0.7353**
	S1 Validation	0.967	y = 1.067x - 0.090	1.3008	0.01077	0.8279	
	S1, Testing Testing	0.963	y = 0.879x + 0.154	1.4016	0.01914	1.3655	
	S2 set, Training Training	**0.991**	**y = 1.004x - 0.008**	**1.3654**	**0.002642**	**0.1934**	**0.3865**
	S2 validation	0.988	y = 0.984x + 0.028	1.3888	0.007015	0.5051	
	S2, testing testing	0.979	y = 1.006x - 0.006	1.4232	0.006565	0.4612	

Table 6.
Summary of R^2 values of training validation and testing data: 7 N in 1st and 10 N in 2nd L, R_a.

Again experiment has been conducted on D2 steel using WEDM by setting the individual optimum parametric combinations (V_g, F_r, T_{on}, T_{off}, W_f and W_t) as 90 (V), 8 (Lit./min), 1.05 (μS), 190 (μS), 2 (m/min) and 900 (g) respectively and found the values of Ra = 0.9638 (μm) at MRR = 105 (mg/min) (**Table 7**).

6. Conclusion

It has been concluded that the best fitted model (S2) for material removal rate and surface roughness of D2 steel has been achieved by artificial neural network

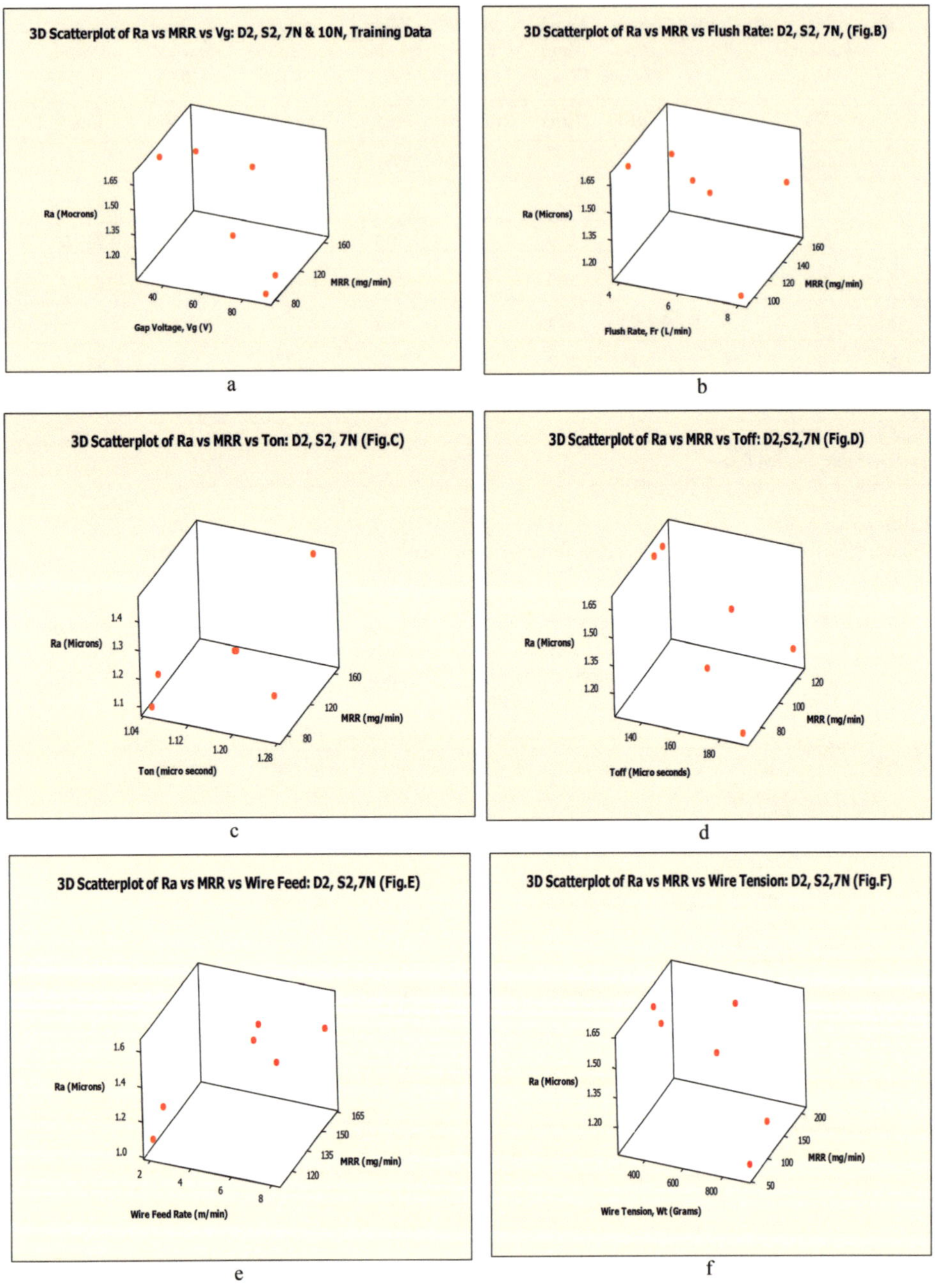

Figure 6.
(a–f) 3D scattered plot between Ra vs. MRR vs. individual independent parameter.

using WEDM. From best modeled training data, optimum parametric combinations (V_g, F_r, T_{on}, T_{off}, W_f and W_t) observed as 90 V, 8 Lit./min, 1.05 μS, 190 μS, 2 m/min and 900 g respectively and found the values of R_a = 0.9638 μm at MRR = 105 mg/min, whereas the average Ra = 1.3654 μm at MRR = 114.8 mg/min. It has been concluded that ANN modeling technique is best fitted for surface roughness prediction and able to successfully minimize (SR) is 29.41% with 8.53% decreases the MRR from its average values on D2 steel using BPANN under WEDM. Such combinations may be applied for industrial application, where it is needed.

SN	Gap voltage (V_g)	Flush rate (F_r)	Spark time (T_{ON})	Spark time (T_{OFF})	Wire feed (W_f)	Wire tension (W_t)	Surface roughness (R_a) obs.	Surface roughness (R_a) predicted	(Zero residual)^2	Material removal predicted (MRR)
	(V)	(Lit./min)	(μS)	(μS)	(m/min)	(g)	(μm)	(μm)	$(\mu m)^2$	(mg/min)
20	**90**	4	1.15	160	8	600	1.1098	1.1096	4.00E−08	79
43	90	**8**	1.25	130	5	600	1.1551	1.1551	0	94
42	90	8	**1.05**	160	2	600	1.0962	1.0905	3.25E−05	74
54	60	4	1.15	**190**	8	900	1.1871	1.1878	4.90E−07	125
34	90	6	1.25	160	**2**	600	1.0832	1.0812	4.00E−06	111
54	60	4	1.15	190	8	**900**	1.1871	1.1878	4.90E−07	125

Table 7.
Best parametric combination with their possible responses.

Author details

Umesh K. Vates[1]*, N.K. Singh[2], B.P. Sharma[1] and S. Sivarao[3]

1 Department of Mechanical Engineering, ASET, Amity University, Uttar Pradesh, India

2 Indian Institute of Technology (ISM) Dhanbad, India

3 UTeM, Melaka, Malaysia

*Address all correspondence to: u.k.vates@gmail.com

References

[1] Gatto A, Luliano L. Cutting mechanisms and surface features of WEDM metal matrix composite. Journal of Material Processing Technology. 1997;**65**:209-214

[2] Azouzi R, Guillot M. On-line prediction of surface finish and dimensional deviation in turning neural network base sensor fusion. International Journal of Machine Tools and Manufacture. 1997;**37**:1201-1217

[3] Montgomery DC. Design and Analysis of Experiments. Singapore: Wiley; 1991

[4] Ho KH, Newman ST, Rahimifard S, Allen RD. State of the art wire electrical discharge machining (EDM). International Journal of Machine Tools and Manufacture. 2004;**44**:1247-1259

[5] Puertas I, Lues CJ. A study on the machining parameters optimization of electrical discharge machining. Journal of Material Processing Technology. 2003;**143-144**:521-526

[6] Sanchez JA, Lopez de Lacalle LN, Lamikiz A, Bravo U. Dimensional accuracy optimization of multi-stage planetary EDM. International Journal of Machine Tools & Manufacture. 2002; **42**:1643-1648

[7] Mondal D, Pal SK, Saha P. Modelling of electrical discharge machining process using back propagation neural network and multi- objective optimizations using non-dominating sorting genetic algorithm-2nd. Journal of Materials Processing Technology. 2007;**186**:154-162

[8] Panda DK, Bhoi RK. Artificial neural network prediction of material removal rate in electro-discharge machining. Materials and Manufacturing Processes. 2005;**20**:645-672

[9] Fisher RA. Statistical Methods for Research Worker. London: Oliver & Boyd; 1925

[10] Sahoo GB, Ray C, Wade HF. Pesticide prediction in ground water in North Carolina domestic wells using artificial neural networks. Ecological Modelling. 2005;**183**(1):29-46

[11] Kexing S, Jiandong X, Qiming D, et al. Optimization of the processing parameters during internal oxidation of Cu–Al alloy powders using an artificial neural network. Materials & Design. 2005;**26**(4):337-341

[12] Juan-Hua S, Qi-Ming D, Ping L. Simulation of aging process of lead frame copper alloy by an artificial neural network. Transactions of the Nonferrous Metals Society of China. 2003;**13**(6):1419-1423